William Baker Fahnestock

Artificial Somnambulism, Hitherto Called Mesmerism, Or Animal Magnetism

William Baker Fahnestock

Artificial Somnambulism, Hitherto Called Mesmerism, Or Animal Magnetism

ISBN/EAN: 9783337242947

Printed in Europe, USA, Canada, Australia, Japan

Cover: Foto ©berggeist007 / pixelio.de

More available books at **www.hansebooks.com**

ARTIFICIAL SOMNAMBULISM.

HITHERTO CALLED

MESMERISM;

OR,

ANIMAL MAGNETISM.

CONTAINING A BRIEF

HISTORICAL SURVEY OF MESMER'S OPERATIONS, AND THE EXAMINATION OF THE SAME BY THE FRENCH COMMISSIONERS.

PHRENO-SOMNAMBULISM; OR, THE EXPOSITION OF PHRENO-MAGNETISM AND NEUROLOGY. A NEW VIEW AND DIVISION OF THE PHRENOLOGICAL ORGANS INTO FUNCTIONS, WITH DESCRIPTIONS OF THEIR NATURE AND QUALITIES, ETC., IN THE SENSES AND FACULTIES; AND A FULL AND ACCURATE DESCRIPTION OF THE VARIOUS PHENOMENA BELONGING TO THIS STATE; INCLUDING ITS DIVISION INTO TWO DISTINCT CONDITIONS, VIZ.: THE WAKING AND SLEEPING, WITH PRACTICAL INSTRUCTIONS HOW TO ENTER AND AWAKE FROM EITHER.

THE IDENTITY OF THESE CONDITIONS WITH OTHER STATES AND MYSTERIES.

TOGETHER WITH

AN ACCOUNT OF SEVERAL OBSTETRICAL CASES DELIVERED WHILE IN THIS STATE; THE PROPER METHOD OF PREPARING SUBJECTS FOR SURGICAL OPERATIONS; THEIR MANAGEMENT DURING AND AFTER THE SAME, AND THE LATEST AND BEST METHOD OF CURING DISEASES, ETC., IN THOSE PERSONS WHO ARE IN THAT CONDITION.

BY

WM. BAKER FAHNESTOCK, M.D.

PHILADELPHIA:
PUBLISHED BY BARCLAY & CO.
610 ARCH STREET.

PREFACE.

MUCH has already been written upon this interesting subject, both by foreign and domestic writers, and the appearance of a new work, by an American author, may at this time be thought unnecessary.

The subject, however, I conceive, has heretofore been but imperfectly understood, and so often misrepresented, even by its most strenuous supporters, that a work giving the facts is loudly called for.

It cannot be denied that the friends of the science have but too frequently overstepped the mark, by mistaking appearances for facts, and relating marvellous powers and phenomena, which, in reality, were but creatures of their own fancy.

On the other hand, the enemies of the science have, unhappily, without any positive or practical knowledge of it themselves, wilfully magnified these misrepresentations, and by torrents of unsparing invective, hoped to crush it at a blow ; erroneously supposing that ridicule could put down or annihilate phenomena which they could not explain by an imperfect philosophy.

It is no wonder, therefore, that so many have stood aloof, or set their faces against it, and at this late day one must almost be ashamed to advocate its truth or demonstrate its usefulness.

This state of things is much to be regretted, and the religious prejudices which have in many instances been excited against it, are much to be deplored.

It has always been a matter of astonishment to me to see and hear those who profess to be so much in favor of doing good, blindly opposed to this science, which of all others is most likely to bring health to the afflicted and joy to the sore at heart. A want of knowledge in regard to its true nature has been the cause of this much-to-be-regretted neglect or oversight.

The time for reformation, however, has now come, and it is to be hoped that Somnambulism will soon be entirely rescued from the hands of charlatans, and placed upon a footing with the more favored branches of knowledge.

Scientifically applied to the various uses which its phenomena warrant, it will soon be inseparably joined to Medicine and Surgery, and with those branches be co-equal in relieving disease.

Experience has already sufficiently tested its usefulness, not only in surgical operations, but in a long catalogue of diseases, etc., upon which protracted courses of medicine had no beneficial effects.

This I know to be true, yet it is necessary for every one to see before they can believe. I am content that every one shall enjoy that privilege, and should be very sorry to censure those who do not believe, or credit that which has not been made evident to their senses.

Every one has a right to doubt, but they should not hastily condemn phenomena, or assert that they are not true, until they can positively and unequivocally prove the contrary.

With these remarks, I submit this work to the public, with a perfect conviction that its contents are true, and with hopes that before any portion of it is approved or condemned, that it be thoroughly and scientifically investigated.

CONTENTS.

CHAPTER I.
HISTORICAL SURVEY.

Mesmer not the discoverer of the state—His theory of it—Its examination by the French commissioners—Their conclusions—The author's remarks 49

CHAPTER. II.

Of the causes which have retarded the progress of the science 63

CHAPTER III.

Of the conditions necessary for the production of the somnambulic state, with instructions how to enter it, etc 67
   I.—Of the instructor or "operator." 67
   II.—Of the patient 68
   III.—Instructions 69
   IV.—Of the sensations experienced by those who enter this state 72
   V.—Of their awaking 73

CHAPTER IV.

Theory of this state 75

CHAPTER V.

Of the somnambulic proper sleep 81
   I.—Of a partial state of Artificial Somnambulism 82

CHAPTER VI.

Phreno-Somnambulism 85

CHAPTER VII.

Of the senses.. 94
   I.—Motion; or, the power to move............................ 94

CHAPTER VIII.

Of the functions of the faculties.................................. 96
   I.—Consciousness 97
   II.—Attention 98
   III.—Perception.. 98
   IV.—Memory 100
   V.—Association... 100
   VI. AND VII.—Likes and Dislikes............................. 100
   VIII.—Judgment... 101
   IX.—Imagination 103
   X.—Will 105

CHAPTER IX.

Of the peculiar functions of perception in the different faculties while in a natural state................................. 109
   I.—Of the peculiar functions of perception when in a state of Artificial Somnambulism............................ 115
   II.—The functions considered when in a state of Artificial Somnambulism.. 117
      1.—Consciousness 117
      2.—Attention 117
      3.—Perception 119
      4.—Memory 121
      5.—Association... 122
      6 and 7.—Likes and Dislikes................................ 123
      8.—Judgment 123
      9.—Imagination 123
      10.—Will.. 125

CHAPTER X.

Of reading or knowing the mind................................... 127
   I.—Illustration 127
   II.—Illustration 129
      Theory of Dr. Collyer 135
      Mental alchemy or electrifying............................. 138

CHAPTER XI.

   I.—Of the identity of other mysteries with this state.... 141

II.—Of the mysteries practiced by the modern magicians of Egypt.. 144
III.—Of the "mysterious lady."............................... 157
IV.—Of the earth mirrors....................................... 157
    First earth glass.. 159
    Second earth glass....................................... 160
V.—Second sight.. 161
VI.—Phantasms... 162

CHAPTER XII.
Transposition of the senses...................................... 171

CHAPTER XIII.
Natural sleep... 182

CHAPTER XIV.
Natural Somnambulism... 183
  I.—Trance.. 185

CHAPTER XV.
Of intuition... 193

CHAPTER XVI.
Presentiment or foreknowledge................................ 198

CHAPTER XVII.
I.—Of interior prevision.. 202
II.—Of exterior prevision....................................... 204
III.—Prophetic dreams... 208
IV.—Witchcraft.. 208

CHAPTER XVIII.
Sympathy.. 210
  I.—Clairvoyance.. 215
    Clairvoyance at a distance........................... 221

CHAPTER XIX.
Of the sense of hearing... 237

CHAPTER XX.
Of the senses of smell and taste............................... 242

CHAPTER XXI.

Of the sense of feeling.. 250

CHAPTER XXII.

Of the sense of motion... 258
Of their physical strength.. 261

CHAPTER XXIII.

Of the influence of Artificial Somnambulism on the system. 262
    I.—Of its influence upon a healthy subject................. 262
    II.—Of the influence of Artificial Somnambulism upon diseased subjects... 266

CHAPTER XXIV.

Artificial Somnambulism considered as a therapeutic agent. 269

CHAPTER XXV.

Of the kinds of disease cured while in this state............... 287
    I.—Chorea, or St. Vitus's dance.................................. 288
    II.—Epilepsy... 290
    III.—Dyspepsia.. 290
    IV.—Intermittent fever... 290
    V.—Fever.. 291
    VI.—Case.. 293
    VII.—Inflammatory rheumatism............................... 294
    VIII.—Chronic rheumatism..................................... 295
    IX.—Hysteria... 296
    X.—Melancholy from unrequited love..................... 296
    XI.—Case... 297
    XII.—Case.. 297
    XIII.—Case... 298
    XIV.—Contraction of the muscles of the fingers......... 299
    XV.—Scarlet fever... 300
    XVI.—Case.. 301
    XVII.—Case.. 306

CHAPTER XXVI.

Surgical operations... 313

CHAPTER XXVII.

Obstetrical cases... 315
Conclusion... 316

ARTIFICIAL SOMNAMBULISM.

CHAPTER I.

HISTORICAL SURVEY.

MESMER'S THEORY—ITS EXAMINATION BY THE FRENCH COMMISSIONERS—THEIR CONCLUSIONS—THE AUTHOR'S REMARKS, ETC.

IT is now well known that Mesmer was not the first who discovered what he was pleased to denominate Animal Magnetism. Its application to the cure of diseases had been practised by the Gauls, Romans, Greeks, Hebrews, and even the Egyptians; but without attempting to trace it back to the more remote periods of antiquity, it will be sufficient to remark, that early in the seventeenth century, a celebrated physician, Van Helmot, exhibited a knowledge of it in his writings; and William Maxwell, an Englishman, in sixteen hundred and seventy-nine (1679), laid down propositions similar to those promulgated by Mesmer.

About the middle of the same century, several other operators appeared in England—a Doctor Streper, Leveret, and Valentine Greatrakes, who professed to cure diseases by stroking with the hands.

Greatrakes was quite celebrated at that time, and is said to have performed many cures, which were authenticated by the Lord Bishop of Derry, and many other respectable individuals.

The Royal Society also examined into the mystery, and accounted for the phenomena by supposing that there existed a "sanative contagion in Mr. Greatrakes' body, which had an antipathy to some particular diseases, and not to others."

Many other speculations of the kind were at that time proclaimed and advocated, which it is not necessary to notice. Without, therefore, referring to any others anterior to the time of Mesmer, I shall now give a brief account of his operations.

Although Mesmer was not the first who discovered or applied Somnambulism, or Animal Magnetism, to the cure of diseases, to him is undoubtedly due the credit of its revival.

His first attempts were made in seventeen hundred and seventy-three (1773), and soon after, the artful misrepresentations of Father Hell and Ingenhouse brought it into disrepute; and Mesmer, almost in despair, left his native country, and arrived in France in seventeen hundred and seventy-eight (1778), where, in consequence of the extent to which it was carried on in Paris, the French king appointed a committee, consisting of four physicians, and five members of the Royal Academy of Sciences, to investigate the matter, in the year seventeen hundred and eighty-four (1784).

The former were Borie, Sallin, d'Arcet, and Guillotin; the latter, MM. Bailly, Leroi, de Bory, Lavoisier, and Dr. Benjamin Franklin, then the American Minister at Paris.

Unfortunately for the science, most of the medical gentlemen selected to investigate the facts and the pretentions of the new doctrine, had pre-judged the question, and like too many of the faculty of our own time, were resolved not to be convinced. It is true, the practice of Mesmer was quackery in the extreme, for after refusing to sell his secret to the French government, which had negotiated with him for the purchase of it, he sold it to individuals, requiring their secrecy, at the rate of one hundred louis a head.

Secrecy, however, was not long maintained, and a knowledge of the science was soon propagated and widely diffused, with many additions and corruptions added by various individuals, according to their respective fancies or marvellous propensities.

His theory, however, as published, is as follows: He affirmed that the "Animal Magnetic sleep" (somnambulic sleep) was produced by "a fluid universally diffused, and filling all space, being the medium of a reciprocal influence between the celestial bodies, the earth, and living beings; it insinuated itself into the substance of the nerves, upon which, therefore, it had a direct operation; it was capable of being communicated from one body to other bodies, both animate and inanimate, and that at a considerable distance, without the assistance of any intermediate substance; and it exhibited in the human body some properties analogous to those of loadstone—especially its two poles. "This—Animal Magnetism," he added, "was capable of curing directly all the disorders of the nervous system, and indirectly other maladies; it rendered perfect the operation of medicines, and excited and directed the salutary *crisis* in the power of the physician. Moreover it enabled him to ascertain the state of health of each individual, and to form a correct judgment as to the origin, nature, and progress of the most complicated diseases, etc."

Mons. D'Eslon, who was a pupil of Mesmer, also practised "Animal Magnetism" at Paris, and undertook to demonstrate its existence and properties to the commissioners, and read a memoir, in which he maintained that: "there is but one nature, one disease, and one remedy, and that remedy is Animal Magnetism."

Such were the principal theories and opinions entertained at that time. The method used by Mesmer and his pupils to induce "*the crisis*" (or sleep) will next occupy our attention.

Many individuals were operated upon by Mesmer and M. D'Eslon, at the same time, and the manner of operating is thus described by the commissioners:

"In the middle of a large room was placed a circular chest of oak, raised about a foot from the floor, which was called the bequet (or tub): the lid of this chest was pierced with a number of holes, through which there issued movable and curved branches of iron. The patients were ranged in several circles round the chest, each at an iron branch, which, by means of its curvature, could be applied directly to the diseased part. A cord, which was passed round their bodies, connected them with one another, and sometimes a second chain of communication was formed by means of the hands, the thumbs of each one's left hand being received and pressed between the forefinger and thumb of the right hand of his neighbor. Moreover, a piano-forte was placed in a corner of the room, on which different airs were played; sound being, according to the principles of Mesmer, a conductor of Magnetism."

"The patients thus ranged were, besides, directly magnetised, by means of the fingers of the magnetiser, and a rod of iron, which he moved about before the face, above or behind the head, and over the diseased parts, always observing the distinction of the magnetic poles, and fixing his countenance upon the individual."

"But, above all, they were magnetised by the application of the hands, and by pressure with the fingers upon the hypochondria, and abdominal regions, which was often continued for a long time, occasionally for several hours together."

"The patients subjected to this treatment at length began to present various appearances in their condition, as the operator proceeded. Some of them were calm and tranquil, and felt nothing; others were affected with coughing and spitting; others again experienced slight pains, partial or universal heats, and considerable perspiration; and others were agitated and tortured with convulsions. These convulsions were extraordinary in their number, severity, and duration; and in some instances lasting for three hours, when they were accompanied with expectoration of a viscid phlegm, which was ejected by violent efforts and sometimes streaked with blood. The convulsions were characterized by violent involuntary motions of the limbs, and of the whole body, by spasms of the throat, by agitations of the epigastrium, and hypochondres, and wandering motions of the eyes, accompanied by piercing shrieks, weeping, immoderate laughter and hiccough."

They were generally preceded or followed by a state of languor and rambling, or a degree of drowsiness or even coma. The least unexpected noise made the patient start, and it was remarked that even a change of measure in the air played upon the piano forte affected them so that a more lively movement increased their agitation, and renewed the violence of their convulsions. All seem to be under the power of the magnetiser; a sign from him, his voice, his look, immediately rouses them from a state even of apparent sopor."

"In truth," added the commissioners, "it was impossible not to recognise in these constant effects, great power or agency, which held the patients under its dominion, and of which the magnetiser appeared to be the sole depository."*

---

* Rapport des Commissaires chargès par le Roy de l' Examen du magnetisme Animal : a' Paris, 1784.

I have extracted the above minute account, not only to give an idea of the proceedings at *that time*, but also to show the effects which an individual may produce by working upon the minds of those who are ignorant of the facts. The convulsions, pains, heats, spitting, coughing, and spasms, or the immoderate laughter, the weeping, shrieking, hiccough, languor and coma, etc., were the effects, not of Mesmer's powers, or of Magnetism, but of the *belief* which he had instilled into the respective individuals, that such effects were necessary. His power over them, too, was for the same reason unlimited. I have seen many similarly affected who entered this state under the care and crude directions of those who practise it at the present day; but I have never seen anything of the kind, when the subjects have had a proper explanation of the nature of the state before they enter it.

Mesmer called the convulsive or lethargic state—the *crisis*—and erroneously considered it necessary for the purpose of curing diseases. But I shall speak of this more fully in another chapter.

"The commissioners remarked, that of the many who fell into the *crisis*, most of them were women; that the *crisis* was not effected in less than one or two hours, and that when one person was thus taken the rest were similarly seized shortly after."

As the commissioners, however, were unable to obtain satisfactory results by experimenting upon so many at once, they resolved to experiment upon individuals in a state of health, and submitted to the process themselves, for three days successively, without any effect being produced upon any of them. They therefore concluded that "magnetism has no agency in a state of health, or even in a state of slight indisposition."

Their next trial was upon persons actually diseased; and of fourteen individuals, five experienced some effects from the operation, but nine felt none whatever. Of the five who experienced it, three were ignorant and poor. The commissioners remarked, at the same time, that children, and those who were better able to observe and describe their sensations, felt nothing. They therefore thought that the effects might be explained by natural causes, and attributed the result to the *imagination,* and next commenced a new series of experiments, to determine "how far the imagination could influence the sensations, and whether it could be the cause of all the phenomena attributed to magnetism."

The commissioners had recourse now to M. Jumelin, who magnetised in the same way with MM. Mesmer and D'Eslon, except that he made no distinction of the magnetic poles.

Many experiments were made, and although the commissioners were convinced that the imagination was capable of producing pain, and a sense of heat, etc., yet the effects of *Animal Magnetism* appeared to them more severe, and it was yet to be ascertained whether, by influencing the imagination, convulsions or a complete *crisis* witnessed at the public treatment, could be produced.

To test this point, many experiments were instituted by the commissioners, the result of which seemed to convince them that the *imagination* and *imitation* produced precisely the same effects, and that their experiments were altogether adverse to the principles of magnetism, not negatively, but positively and directly.

That the nature and extent of these experiments may be better understood by those who have not seen an account of them, I will give a few of them in detail, with such remarks of my own as the facts may render necessary.

The magnetisers of that day had affirmed "that when a tree or even an inanimate substance had been touched by them and charged with magnetism, every person who stopped near the tree would feel the effects of this agent, and either fall into a swoon or into convulsions."

"Accordingly, in Doctor Franklin's garden at Passy, an apricot tree was selected, which stood sufficiently distant from the others, and was well adapted for retaining the magnetism communicated to it. M. D'Eslon, having brought thither a young patient of twelve years of age, was shown the tree, which he magnetised, while the patient remained in the house, under the observation of another person. It was wished that M. D'Eslon should be absent during the experiment; but he affirmed that it might fail, if he did not direct his looks and his cane towards the tree. The young man was brought out, with a bandage over his eyes, and successively led to four trees, which were *not magnetised*, and was directed to embrace each during two minutes; M. D'Eslon, at the same time, standing off a considerable distance, and pointing his cane to the tree actually magnetised."

"At the first tree the young patient, upon being questioned, declared that he sweated profusely; he coughed and expectorated, and said that he felt pain in the head: he was still about twenty-seven feet from the magnetised tree. At a second tree, he found himself giddy, with the headache: he was now thirty feet from the magnetised tree. At the third the giddiness and headache were much increased; he said he believed he was approaching the magnetised tree; but was still twenty-eight feet from it. At length, when brought to the fourth tree, *not magnetised*, and at a distance of twenty-four feet from that which was, the *crisis* came on; the young man fell down, in a state of insensibility, his limbs became rigid, he was carried to a grass field, where M. D'Eslon went to his assistance and restored him."

In another experiment "M. D'Eslon was requested to select from among his poor patients, those who had shown the greatest sensibility to the magnetism; and he accordingly brought two women to Passy. While he was magnetising—Doctor Franklin and several persons in another apartment—the two women were put in separate rooms. Three of the commissioners remained with one of the women, the first to question her, the second to write and the third to represent M. D'Eslon, who (they persuaded her, after bandaging her eyes) was brought into the room to magnetise her. One of them pretended to speak to M. D'Eslon, requesting him to begin; *but nothing was done;* the commissioners remained quiet, only observing the woman."

"In the space of three minutes she began to feel a nervous shivering (frisson nerveux); then she felt in succession, a pain in the head and in the arms, and a pricking in the hands; she became stiff, struck her hands together, got up from her seat, and stamped with her feet—in a word the *crisis* was completely characterised."

"Two of the commissioners were in an adjoining room with the other woman, whom they placed by the door, which was shut, with her eyes at liberty, and made her believe that M. D'Eslon was on the other side of the door, magnetising her. She had scarcely been seated a minute before the door, when a shivering began; in one minute more she had a clattering of the teeth, but yet a general warmth over the body; and by the end of three minutes the *crisis* was complete. The breathing became hurried, she stretched out her arms behind her back, writhing them strongly, and bending the body forward; a general tremor of the whole body came on, the clattering of the teeth was so loud as to be heard out of the room, and she bit her hand so as to leave the marks of her teeth in it."

With respect to the first experiment, the commissioners remarked, "that if the patient had experienced no effects under the tree actually magnetised, it might have been supposed that he was not in a state of sufficient susceptibility; but he fell into the *crisis* under one which was not magnetised; therefore, not from any external physical cause, but solely from the influence of the *imagination*. He knew that he was to be carried to the magnetised tree; his imagination was roused, and successively exalted, until at the fourth tree it had risen to the pitch necessary to bring on the *crisis*."

This reasoning, at first sight, seems very natural and conclusive; but with due deference, I must say that it is very far from the truth; and the experiments only prove that *magnetism is not* the cause of this state, and nothing more.

I contend that it was not the imagination (in the strict sense of the word) which threw the above persons into that condition.

The young man, who was led blindfolded to the different trees, was made to *believe* that if he embraced a tree, or anything else which was, *as they said*, magnetised, he would fall into this state. What was the result? He fell into it before he reached the tree which was said to be so operated upon. Yet this does not prove conclusively that it was his imagination which produced that effect.

I have instituted many experiments to determine the cause of this condition, and all the facts gathered go to prove that the state can be entered by an act of the subject's own will, or can be induced by the *belief* (on the part of the subject) that another person has the power of throwing him into it. Therefore, in the case of the boy above alluded to, the *belief*, or even the suspicion, on his part, that the tree (*whether magnetised or not*) would have such effects, was sufficient to make him enter it—he not knowing that he could resist or enter it at pleasure; consequently he fell into it, as a matter of course, at the tree, *which he thought* or *believed to be* magnetised.

The same was the case with the two women. But if they had been acquainted with the true nature of the state, and their powers to resist, etc., neither the tree, the operator, nor the deceptions practised, would have had any effect upon them; and if the commissioners had known that it was in the power of these individuals to throw themselves into this state *at will*, independent of any person or any foreign cause, their conclusions would have been very different.

The experiments of the commissioners prove, most conclusively, that magnetism has no agency in producing this condition—indeed, that there is no such thing in nature as animal magnetism, and that consequently it cannot be communicated to anything else, and that the effects ascribed to it were absolutely the result of a *belief* that certain effects would follow certain operations, and not the effects of any foreign cause; and until the miserable *belief* in a magnetic fluid is abandoned or eradicated from the mind of man, we will have unnecessary suffering, and the world will be deprived of the superior benefits which a properly directed mind (while in a somnambulic condition) would have upon disease.

This is the grand secret of curing diseases, and all that has ever been effected by entering this condition, has been effected by the mind of the *subject* while in that state—not knowingly, however, but in many cases by bringing the mind accidentally to bear upon it. How much more ought we to expect under the proper management of the mind while in that condition? The sooner, therefore, that we do away with all passes and useless operations of every kind, the sooner we will act like rational beings and reap the benefits of a regenerated science.

At the time when Mesmer revived the science the magnetisers, as well as the world at large, knew very little about the science, or the extraordinary phenomena, powers, etc., belonging or peculiar to persons in that state, and their experiments were made to ascertain its cause before they knew its effects. This has, unfortunately for the science, been the case with most experimenters from that time until the present day, and many other theories, equally untrue and unfounded, have, phœnix-like, arisen and are still supported by their respective adherents. It is not my intention, however, to examine any more of them particularly, as the facts, which I shall present to the reader in this work, will render it unnecessary.

CHAPTER II.

OF THE CAUSES WHICH HAVE RETARDED THE PROGRESS OF THE SCIENCE.

THE causes which have retarded the progress of the science are very numerous, and among the number which stand most conspicuous are the many misrepresentations which have been made, both by its friends and its foes.

It is not to be denied that its foes have in many instances stooped to falsehood and wilful misrepresentations, while it is also true that its friends, in their enthusiasm, reported appearances as facts, no matter how contradictory they were in themselves, or how unreasonable they might appear to those who would scrutinize them with different views.

It is chiefly to these causes that I ascribe most of the difficulties which have arisen, and the torrents of ridicule which have been at different periods showered upon the devoted heads of those who have advocated the science.

I am aware that this has been the case with all the other sciences already established, but is it the best, most correct and speedy method of seeking for the truth? I think not; and it seems to me that before a *postulatum* is condemned, it ought to be thoroughly examined; and that in the investigation of a new science, it is absolutely necessary first to make ourselves acquainted with the facts, and to examine them individually and collectively with the utmost care, before we can arrive at conclusions with that degree of certainty which the truth, in its might, always renders self-evident, and leads us, as it were unconsciously, to that system which the God of Nature has established from the foundation of the world.

Had investigators, generally, confined themselves strictly to facts, instead of framing theories as visionary as they were often false, there would have been less confusion and fewer marks for the shafts of ridicule; and the sciences, instead of being retarded, would have flourished at every step, and mankind much sooner reaped the benefits which they were destined to confer.

These remarks apply themselves particularly to the science of Artificial Somnambulism, which of all others, is perhaps the most prolific in phenomena, of so mysterious and unaccountable a nature, that the study of its peculiarities has been rendered doubly difficult, not only on account of numbers, but the apparent contradictions which have been exhibited by different subjects, or the same subjects at different times, together with the obscurity which the ignorant and superstitious have thrown around it, rendering it still more difficult to distinguish those facts which alone could lead to correct conclusions. Many things, too, have been laid at the door of this science which do not belong to it, and respectively have also created many difficulties which will have to be refuted before the science can be brought before the world in its true light, and the community induced to look upon it with that consideration which it truly deserves.

It is not an easy matter for skeptical persons to get the necessary proofs to convince them of the truth of the science, as subjects, who enter this state under the care of another, cannot often be persuaded to perform experiments to please those whom they may not even take an interest in pleasing when awake, much less so when they are in this state, with all their powers active; particularly their capability of reading the mind, by which they learn the motives which *actuate* and the doubts which exist and are naturally always uppermost in their minds. They cannot, in the face of all this, be persuaded to give such persons satisfaction.

I am aware that skeptics do not believe that they have such powers, but their unbelief does not make it the less true; and I can assure them that they never will get satisfaction unless they experiment themselves, and throw aside all prejudice and investigate the matter, as they would any other science, purely for the love of truth.

As all the phenomena are extraordinary in their nature, they must be seen to be believed. It is therefore impossible for one man to convince every one, and although I have convinced thousands, there are still thousands who disbelieve, because they have not had the same opportunities and explanations. I have been hooted at, decried as a charlatan, a wizzard, and a fool—yet day after day I have also had the satisfaction of seeing these very wise opposers convinced of its truth, and obliged to ask for the very knowledge which at one time they seemed so heartily to despise. Such is the power of truth, which, sooner or later, must prevail.

I have been told by some persons that they had the utmost horror of the state, and utterly despised the science. And why? They could give no reason or even say in what particular it had offended their delicacy. They were simply prejudiced—utterly ignorant of its nature and wilfully blind to its benefits, and opposed it because it was something they did not understand.

Every man has a right to doubt, and I yield every one that privilege; but we have no right to anticipate and form prejudices before we have examined the matter for ourselves.

Every new science is liable to misrepresentation, and, although we may not at first be able to understand it, we should not unhesitatingly say that it is untrue because we cannot fathom it at a glance, or reconcile it to our previous notions. Every thing is plain and palpable when properly understood, and we should rather always patiently await its elucidation, than hazard an opinion, or run the risk, not only of making ourselves ridiculous, but of retarding, by our opinions, the progress of a science which in time may become useful.

CHAPTER III.

OF THE CONDITIONS NECESSARY FOR THE PRODUCTION OF THE ARTIFICIAL SOMNAMBULIC STATE.

MUCH has been said about the conditions necessary to produce this state, and I am sorry that I must differ from all that has been written upon this subject.

I.—OF THE INSTRUCTOR OR OPERATOR.

From what has been said, the reader must already know that the "operator" has no power to produce this condition; and independent of his instructions, and his capability of managing them while in it, has nothing to do with it.

His health, temperament, age, etc., as a matter of course, is also immaterial, so that his intelligence, mental character, and knowledge be of such a nature as to be worthy of the trust placed in him.

His skill in managing persons and curing diseases, etc., will depend entirely upon his knowledge of the state, his acquaintance with the nature of diseases, and his intelligence and tact in fixing and properly directing the minds of his patients.

Physicians are best calculated for this purpose, and should always be preferred if they have made the subject their study.

II.—OF THE PATIENT.

I have never found any perceptible difference in what has been called the susceptibility of persons of different temperaments, and have had as many of the Bilious, Bilious Nervous, or Bilious Nervous Sanguine to enter it, as of any of the other temperaments.

Too much stress has been laid upon this subject by those who have heretofore written upon it, no doubt from an anxiety to account for the failures which they often experienced; and I again repeat that I have found no material difference in this respect which would induce me to believe that it was produced by a difference in the temperaments, nor have I observed much difference between the readiness with which it is entered by the different sexes. I have found some men of opposite temperaments to enter this state more readily than some women of the same temperaments, and *vice versa*, and believe that what is termed susceptibility, or a readiness to enter it, depends *more* upon the *state of the subject's mind at the time of trial*, than upon sex, temperament, or phrenological developments, etc.

Noise, being afraid of it, an over anxiety to enter it, risibility, and in fact any other mental excitement, is unfavorable to its accomplishment, and should always be avoided as much as possible.

Very old persons, and children under eight or ten years, from a want of sufficient steadiness, knowledge and determination, cannot often be induced to enter it perfectly.

III.—INSTRUCTIONS.

Various methods have been employed by different operators to induce this state. The plan adopted and practised by Mesmer and his pupils has already been detailed in a preceding chapter of this work. Those of modern magnetizers are scarcely less absurd than that employed by Mesmer and his immediate followers.

Some operators of the present day, who believe in a magnetic influence, still pursue the ludicrous method of sitting down opposite to the patient, holding his thumbs, staring into his eyes, and making passes, etc., until the desired object is effected.

Others, who believe looking to be essential, direct the patient to look at some object intently until the lids close and the patient becomes unconscious.

Very few, however, can be induced to enter the state in any of the above ways, and those who do, usually fall into the *sleeping condition* of this state, and are generally dull, listless, and seldom good clairvoyants.

The most rational, certain, and pleasant way of inducing this state which I have discovered, is the following:

When persons are desirous of entering this state, I place them upon a chair where they may be at perfect ease. I then request them to close the eyes at once, and to remain perfectly calm at the same time that they let the body lie perfectly still and relaxed. They are next instructed to throw their minds to some familiar place—it matters not where, so that they have been there before and seem desirous of going there again, even in thought. When they have thrown the mind to the place, or upon the desired object, I endeavor, by speaking to them frequently, to keep their mind upon it; viz.: I usually request them to place themselves (in thought) close to the object or person they are endeavoring to see, as if they were really there, and urge them to keep the mind steady, or to form an image or picture of the person or thing in their mind, which they must then endeavor to see. This must be persevered in for some time, and when they tire of one thing, or see nothing, they must be directed to others successively, as above directed, until clairvoyancy is induced. When this has been effected the rest of the senses fall into the state at once or by slow degrees—often one after another, as they are exercised or not—sometimes only one sense is affected during the first sitting. If the attention of the subject is divided, the difficulty of entering the state perfectly is much increased, and the powers of each sense while in this state will be in proportion as that division has been much or little.

Almost every subject requires peculiar management, which can only be learned by experience or a knowledge of their character, etc. Much patience and perseverance is often required to effect it; but if both be sufficiently exercised, the result will always be satisfactory—if not in one sitting, in *two or more*. I have had several to enter this condition after twenty (20) sittings, and had them to say "that if they had not interfered, and let things take their course, they would have fallen into it in the first sitting." This shows that those who do not enter it in one or two sittings, must do something to prevent it.

Many persons have entered the state in the above manner who could not do so in any other, although repeated trials had been made to effect it.

Taking hold of the thumbs and looking into the eye, or at any other object particularly, *is by no means necessary;* and as this state is one that depends entirely upon the state of the subject's mind, and is brought about by an act of his own will and not by that of the operator's, it must be evident to every intelligent mind that all that the operator can do, independent of the instructions which he may give, or the care he may take of them, etc., is perfectly useless and ought to be dispensed with.

I have found that persons always enter this state better without any contact, looking, passes, or any thing of the kind, particularly when they are assured that they have some competent person to take care of and to converse with them while in it; and, by observing carefully the instructions which I have given, it is possible for any person to throw themselves into this state *at pleasure*, independent of any one; but it might not always be prudent to do so for the *first time*, for *some*, upon entering the condition for the *first time*, become unconscious of all that is passing around them; and if such persons were to throw themselves into it independent of any one, and had not consented, or made up their minds before entering it, to hear or to speak to some one, it is most likely that when in it and spoken to, they would not hear any one, and in all probability would sleep for a longer or shorter time, without doing anything, and when they did awake, would remember nothing and scarcely know that they had been in it at all. Or they might get up and wander about, as is sometimes done by natural somnambulists, and unknowingly get into difficulties, or meet with some accident which might not be very agreeable when they awoke.

It is, therefore, always better for those who wish to enter it to place themselves under the care of some one; and he who understands the nature of the state best, and has had the most experience in its management, is the best calculated for this purpose. When they have entered the state frequently, and have had the proper instructions *while in it*, the case is very different; they are then able to move about with as much certainty and safety as if they were awake.

IV.—OF THE SENSATIONS EXPERIENCED BY THOSE WHO ENTER THIS STATE.

The sensations experienced by those who enter this state are variously described by different subjects; but most commonly they agree that after the eyes are closed, and they have been endeavoring to see for a longer or shorter period, a drowsiness ensues, accompanied with more or less "swimming of the head," and a tingling sensation creeping over the whole body.

Some experience a feeling of sinking down as if they were passing through the floor; others, again, feel light as a feather, and seem to ascend or to be suspended in air. Some start and twitch involuntarily in various parts of the body, while in others the breathing is more or less affected, but there is no necessity for their feeling unpleasant in any way. Some feel warm, others cold, but none of the sensations are described as being unpleasant; and when the state is entered perfectly the feelings are said to be delightful.

V.—OF THEIR AWAKENING.

All that is needful, when it becomes necessary that they should awake, is to ask them whether they are ready or willing to do so, and if they are, I direct them to do so at once, and they will awake at the word Now! in an instant.

If, however, you should desire to awaken them, and they are not willing, it will be found impossible to do so contrary to their will, and you will be obliged to await their pleasure.

Before they awake, however, I commonly request them to remember how they felt and what they saw, etc., or they may not know anything about it when they do awake; particularly if it be their first sitting. With some this is not necessary after the first or second sitting, as they commonly make up their minds to do so of their own accord. Yet I have seen some with whom it was always necessary; indeed, I have two subjects with whom I have the greatest difficulty when asleep to persuade them to remember anything. Yet when they awake they are much mortified at not having any recollection of what has transpired; and it seems as if it were impossible for them to carry the resolution to do so *into* that state, and when in it, to resolve to *remember* when they awake.

Others, on the contrary, have the power of remembering whatever they please, or of forgetting what they please; or, in other words, they can remember all that has transpired, only a part of it, or nothing at all, as they may feel disposed at the time.

This quality or power of the mind while in this condition, enables them to *create* pain or feel pleasant at will, and if they imagine, or determine, that there is, or shall be pain or disease in any part of the body, that pain or disease will certainly be felt, at the time and place designated, and will continue until the mind acts, or is directed so as to alter the condition. This peculiar power of the mind while in this state, I have taken advantage of to cure diseases, and if the mind be properly directed while in this state, so as to make them resolve to be well, pains, contracted habits or diseases are removed by an act of their will, as if by magic, and will last until the conditions are changed or altered by influencing causes, or by a positive act of the subject's will.

CHAPTER IV.

THEORY OF THIS STATE.

THE agent or cause of this state (as I have before intimated) has been ascribed by different writers to Animal Magnetism, Mineral Magnetism, Electricity, Galvanism, a Nervous Fluid, and finally to Nervous Induction or Sympathy, etc. But as I have before stated that it is not my intention to examine any more of the theories in detail, I shall here only (as it is the latest) briefly notice that of Sympathy.

In a letter addressed to the Editor of the "Lancaster Intelligencer and Journal," dated October sixteenth, eighteen hundred and forty-three (1843), and published in that paper the following week, I gave my views respecting the agent, in (what has been called) "Animal Magnetism;" and after expressing that I was fully aware that they differed from every other theory extant, and trusting that they would be as near the truth as reasoning from the facts collectively would admit, I go on to say, in refutation of the theory of Sympathy, that: "I have long since proved that when a person enters this state, he does so independent of the operator, his passes, or his will, and while in this state is independent of him, in every sense of the word, and, if he feels disposed, can resist him in every possible way. He can feel different, have different tastes, hold other views, *and it is only by consent that the subject hears him at all, or does anything he requests.* It is true, that some subjects *do apparently* exhibit signs of sympathy with the operator, and will taste what he tastes, feel what he feels, or even think as he will, etc., but they can do the same things with any other person as well as with the operator, although he may, at the same time, will them to do the contrary as much as he pleases. If they do not wish to perform an experiment, the operator has as little influence over them as any other person, and if they are not paying attention to him, it is always necessary for him to draw their attention before he can succeed in any of his experiments.

If this state were induced by sympathy, or they were more in sympathy with him than with any person else, this would not be necessary, as they would then always be obliged to think, feel, taste, etc., as he did, or as he willed them. It is not sympathy, therefore, which influences or enables them to taste, feel, or know what the operator or other persons are tasting, smelling, or willing, etc., but is owing to a peculiar power which they possess while in the state of *translating their faculties at will*, or of reading the mind, not only of the operator, but of any person else, no matter whether they be in the room or not at the time."

"Again—the author of the sympathetic theory claims the power of Magnetising or paralyzing arms, etc., and ascribes the power to the same cause, viz.: Sympathy."

"I *have apparently* done the same thing years ago—but it was not by Sympathy, or any other influence or power exercised, derived, or, emanating from me, for I have seen subjects (while perfectly awake) paralyze their own arms to perfection, without the aid of any operator at all. They not only put their own arms to sleep, but awake them at pleasure, in part or the whole at once, according to the nature of the experiment, which they wished to perform. Where, I would ask, was the Sympathy derived from in these cases?"

In concluding the same letter, I remarked: "That this state was a peculiar one (*Somnus a voluntate*) independent of Magnetism, Electricity, Galvanism, a Nervous Fluid, Sympathy, or anything of the kind, and was entered by the subject at pleasure. That it was a state into which any person could throw themselves, and awaken themselves, either in part or the whole body at once, slowly or otherwise, independent of any one else, or subject to any one's control."

In a second letter to the editor, on the same subject, dated November ninth, eighteen hundred and forty-three (1843), I stated, in addition to what I said in my first letter, that: "I considered this state similar to Somnambulism: Somnambulism being the natural state, and this one the artificial—effected by the motives or will of the subject."

In a letter upon the same subject, to the Editor of the Philadelphia "Spirit of the Times," dated November twenty-eight, eighteen hundred and forty-three (1843), and published in that paper on the second of December following, I stated, in speaking of the lectures at Lowell, by the author of the sympathetic theory, that: "I am sorry to see that this talented gentleman has again let his imagination run away with his better judgment, and I can assure you that his very learned and beautiful theory of sympathy, is as far from the truth as that which he has just deserted.

Were you with me a few days I could prove this to your perfect satisfaction, but as I can scarcely hope for this pleasure, I must be content to give you a brief statement of facts: I have had over three hundred different individuals to enter this state under my care, and have found by innumerable experiments that they are entirely independent of me, and can enter this state and awaken themselves whenever they please, notwithstanding all I can do to the contrary.

They can throw the whole or any part of the body into this state at pleasure, and I have seen many do it in an instant, or before it would be possible for you to enumerate ten. I have had them to throw in a single finger, a hand, an arm, the whole brain, or even a single organ (or portion), and awake them at pleasure." After giving an account of some new experiments on the organs of the brain, etc., which I shall notice in another chapter, I go on to say that: "I have lately had all those subjects, who before gave no evidence to the touch, to respond to the organs, by simply directing them to throw their minds upon those portions of brain which I designated. This, of itself, is sufficient to refute the doctrines of Magnetism, Neurology and Sympathy."

"The author of the sympathetic theory has deceived himself, and his experiments are calculated to deceive any person who is not acquainted with the facts. *Belief*, or even a *suspicion*, that an operator has or have an influence over them, is often sufficient to make persons ignorant of the facts susceptible, and if they do not resist it, they will fall into it *of their own accord*, as easily as into a natural sleep."

"I have had many to fall into this sleep—and some who were seemingly determined not to do so—by simply stating that at a certain time I would magnetise all in the room, although I was thinking of other things, and did nothing but walk up and down."

"This is the same that the author of the sympathetic theory has done on a large scale, and any person can do the same thing, if he can make the subjects *believe*, or even imagine, that they have the power to produce such effects. All that is necessary is to establish the *belief* that our operator has complete control over him, and that he must sympathise with him when asleep, and the operator will have this power over him; but, let the same subject know the facts, before he enters the state, and the operator will soon find that his powers are airy nothings, and that he was before only obeyed, because the subject blindly consented. They can cast off the operator entirely, and prefer another person at pleasure, independent of any one."

"I am sorry that I am obliged to rob the many scientific gentlemen who are engaged in this science, of their imaginary powers, but it is high time that the veil should be drawn, and the mystery or witchery of the matter done away."

The facts, therefore, warrant and bear me out in saying, that this state is nothing more than artificial somnambulism effected by the motives or will of the subject, and not by any foreign cause, power, influence, or sympathy whatever.

CHAPTER V.

OF A SOMNAMBULIC PROPER SLEEP.

THE somnambulic state has also its own sleep or condition of perfect rest, in which all the faculties are sometimes wrapped, and I have frequently been obliged to indulge subjects in what they call "sleep," while in the waking condition of this state. They seem to be able to fall into this unconscious sleep at pleasure, and frequently have to be aroused, as out of a natural sleep. When aroused, however, they awake into the somnambulic and not into the natural state.

It is into this species of the somnambulic state, that most subjects fall when they are not spoken to, or disturbed, before they are known to have entered the state, particularly when it is their first sitting, and out of which they are aroused, when addressed by the person into whose care they have entrusted themselves.

It is into this state also which they occasionally fall, when they have been too much worried by experiments, and is the unconscious state which I have spoken of as their sleep.

It is evident, therefore, that this state has also two conditions, viz.: A waking state and a sleeping state.

The former may be entered, without losing or forgetting themselves, and is generally entered first, particularly when the patient has been frequently spoken to while entering it.

The latter cannot be entered without losing or forgetting themselves, and is the state into which men subjects usually fall when not spoken to, while entering it, and out of which they would sooner or later awake, without any knowledge of having been in it, if not spoken to during the sleep.

I.—A PARTIAL STATE OF ARTIFICIAL SOMNAMBULISM.

It is generally expected that all persons who are said to be in this state shall exhibit the same phenomena.

This is true, so far as the state is perfect, but it must be remembered that all do not enter this state perfectly at the first sitting, and that there is such a thing as a partial state, in which only one, two or more of the senses are effected at the same time, while the rest remain in their natural condition, and of course cannot exhibit the peculiar phenomena which they are capable of when such senses are truly in this state.

I have seen many cases in which the eye-lids only were effected; sometimes only one or more of the senses, etc.

Many remarkable cases have come under my observation, one of which I will here relate:

Miss ——— had entered this state, but imperfectly

several times, and her appearance, while in this semi-state, was such that one would be induced to believe that it was perfect—and while in it, heard no one, (independent of myself,) was insensible to pain, yet with her eyes bandaged, or when I stood behind her, would imitate me, or place her hands and fingers in every possible position which I chose to place mine, but as soon as spoken to by me, would wake, even at the first word addressed to her, remembering nothing that had passed nor aught that she had done.

This case is remarkable for her not being able to remain in the state when spoken to by me. She heard no one else, because her attention was directed entirely to me, and was insensible to pain, because her mind was completely so engaged.

That she saw, or was unable to read my mind, was evident, or she could not have imitated my motions, etc., with her eyes perfectly bandaged. Her awaking, when spoken to by me, seemed to be irresistible, and was so sudden that she could not be persuaded to remember what had taken place before she was perfectly awake. After a proper explanation of the state had been made to her, she entered it perfectly, and is now an excellent clairvoyant, and can be spoken to, etc., without waking until she is ready.

Some subjects are not clairvoyant, although they are perfectly in the state, and their not seeing in such cases is owing to their not knowing how to direct their mind, or their having no disposition to try. I have, however, succeeded in getting many to see who otherwise would not have done so, by persevering until I persuaded them to try, and instructing them to throw their minds to certain places where they were acquainted, or to hunt up certain individuals whom they were most anxious to see.

The reason why certain senses do not enter this state is owing to their not having been given up to it, or, to a natural or constitutional wakefulness, which, however, I am persuaded can be overcome in all cases by perseverance, and a fixed determination on the part of the subjects themselves.

I have seen several subjects who had frequently been in the state on former occasions, that for a time seemed to have lost the art, and could not enter it again, although they had made many trials to effect it.

The reason why they could not enter it on these occasions, was: Because they had something else upon their minds, and were too impatient or anxious to sit down with sufficient calmness to re-enter it. They have all, however, again succeeded by following the proper instructions, and by noticing particularly the manner in which they re-entered it, have since been enabled to enter it at pleasure.

CHAPTER VI.

PHRENO-SOMNAMBULISM.

I USE this term to express the somnambulic state induced by any subject, in one or more of the organs, faculties or functions of his own brain; or the putting to sleep or awaking the various portions of the brain by the subject himself for phrenological purposes, etc.

I have met with some persons who profess the power of exciting any portion of their own brain at pleasure, even in an instant, although they have never been wholly in a somnambulic state, and at the time of an experiment, to a casual observer would seem to be perfectly awake, or in a natural state, and, indeed, are so, immediately before and after the experiment, but at the instant, or during the time of the experiment, the brain, or a certain portion of it, is in the somnambulic state. These subjects can also perform clairvoyant experiments at pleasure, as well as if the whole body were in a somnambulic state— consequently the mind, *or a certain portion of the brain is in the same condition.* These cases, however, are rare, and when met by neurologists or sympathetic operators, furnish good subjects for their impositions; because these subjects, by merely knowing that they have the power themselves, can excite any of their own organs at will, independent of any operator, and therefore have been imposed upon.

Many public and apparently successful demonstrations of exciting the organs by the touch have been given by various operators, to prove the existence of this power in themselves; which I am sorry to find has been generally credited by those who believe in the science of Phrenology. Many, indeed, were made converts upon the very grounds of this belief, and the experiments of Dr. Buchanan upon the *supposed* impressible subject, seems, in most cases, to have warranted their conclusions.

Although a believer in the general principles of Phrenology, I have always been disappointed in my experiments, instituted to test the possibility of exciting the organs of the brain *by the touch*, in persons who were in a state of artificial somnambulism.

It is true, I have had many subjects to respond to some of the organs, but at least two-thirds never responded at all; and in almost all those who did, I could trace the effects to some positive knowledge of Phrenology on their part. Others again would respond to an organ correctly at one time, and hesitatingly or incorrectly at another. Some who were made to believe that the organ of benevolence was situated on the back part of the head; or where the organ of self-esteem is located, in after times, upon touching that point, would respond to the organ of benevolence; and so with all the other organs respectively; they would respond as they were taught, or as the intimation had been originally received. These facts led me to investigate the cause.

I never doubted the action of the organs themselves, or that the proper portion of brain did not act within the skull, when any evidence was outwardly manifested; but I could not reconcile the facts at that time, or tell the reason why an organ did not always act when the proper locality was outwardly touched. I shall not here relate the numerous experiments which I made, or the many difficulties I had to encounter before I arrived at conclusions which were satisfactory to myself. It will be sufficient to say that my experiments upon those who gave no evidence of excitement from the touch, has been very extensive, and I found that almost all, particularly those who were intelligent, responded to the organs by simply directing them to throw their minds upon those portions of the brain which I designated.

This fact explains the whole mystery of the touch or of the nervous fluid, influence, or whatever else it may be called, which was supposed to penetrate the skull and to excite the brain. The truth, therefore, is self-evident, that the excitement was produced by the subjects themselves, and not by the operator. Those subjects who responded to the touch correctly, in the old method of exciting the organs, I am satisfied must have thrown their minds upon those portions of the brain, beneath the fingers of the operator, and *thus* have excited their organs, while the reverse must have been the case with those who gave no evidence of excitement at all. Some, I am also convinced, received their information of what the operator desired from his own mind, and responded accordingly, particularly in those cases where the organ was not touched, and merely pointed at.

These were my first solutions of the difficulty respecting the touch; but I have still another method of proving the falsity of the doctrines of Neurology and Sympathy, etc., and of establishing the general principles which have been promulgated by phrenologists.

I have already stated, that most subjects can be taught to throw a finger, an arm, or the whole body into this state at pleasure. I will now add, that they can also throw in the whole brain—the half thereof, or even a single organ, or portion, and awaken either at pleasure, independent of any one. This fact which I published in eighteen hundred and forty-three (1843), in the Philadelphia "Spirit of the Times," I have since had many opportunities of confirming. It is true, some have considerable difficulty at first, but when the subject is intelligent and himself anxious to succeed, the power to do so is soon acquired. Some, as I have before stated, are enabled to do so when every other part of the body is perfectly awake; but as a general thing, when the experiments are continued for any length of time, the body also sinks into the condition, but can be aroused by the subject at pleasure, and mostly is so at the close of every experiment.

In my letter to the editor of the paper above

alluded to, and from which I make the following extract, I stated that: "I have had some subjects to throw the whole brain into a somnambulic state, and then successively to awaken one portion of the brain after another, while I noted the results in each case, until the whole brain was in a natural condition.

"These experiments are the most interesting I have ever made; and the results have been the most extraordinary I ever witnessed.

"To give you an idea of these experiments and their results, I will state, that when the brain is in this peculiar state and I request them to awaken an organ, which I point out by stating its position—say Language for example—and then give them (into their hand) some familiar article—a watch, a knife, or a key, etc., it will be impossible for them to name it, as long as that organ is kept awake.

"They know what it is, its use, size, weight, color, and configuration, etc., but they cannot *name* it.

"If I direct them to awaken Tune only, they cannot then distinguish between tones, or recognize the most familiar air, although they can distinguish, know and name other things correctly. If, on the contrary, I direct them to awaken all but a single organ, say that of Tune, then their disposition will be to make or hear tones, etc.

"Every sound is pleasant to them, and even (as a gentleman expressed himself) 'the crackling of the fire seems music'—but if to this organ they add time or harmony, then they become more select, and prefer something more musical.

"When Tune only is in this state, they can judge of nothing but tones—and, although you may have placed fifty different articles in their hands, it will be impossible for them to recognize, name, or tell anything about them; and when they awaken the organ, they will remember nothing that was said or done, except what related to tones, although you may have expressly endeavored to impress other things upon their minds.

"Again: If I request them to throw in the whole brain, with the exception of the organ of Language on the one side, then they will be able to distinguish and name things on one side, and not on the other.

"The experiments may be varied at pleasure, and when the whole brain is in this state, and one organ is awakened after another, as they awaken, they become passive, or lose (in proportion as they are perfectly awake) their will, their power to perceive, imagine, judge, and remember, etc., respectively, until the last is relieved, when they become active at once, and are then in a natural state. If, on the contrary, they are put to sleep, one after another, as soon as the first enters this state, all the rest become inactive at once, and, as they enter successively, they become active until they are all in, and then the brain is in a somnambulic state.

"This proves that the faculties are more active in this than they are in a natural state; and that they possess certain independent or peculiar functions respectively, namely :—Attention, Perception, Imagination, Judgment, Memory, and the Will, etc.—else how could a single faculty when *alone* in this state, remember at will, all that relates to its peculiar functions and *nothing more*, although you may endeavor to impress other things upon the mind of the subject when that faculty was only in this state ?

"In both the natural and somnambulic waking states the functions are awake or ready to act, while in the somnambulic proper sleeps, if they be perfect, they are inactive. But if in either of these proper sleeps one or more of the functions awake, it or they become active, and then dreaming ensues, and we remember what transpires or not, as the memory of the faculty dreaming, is awake or asleep.

"If the function of perception in the organ of Tune, be only in a somnambulic state, the subject will only be able to perceive Tones; but cannot imagine, judge, or remember them, unless these functions respectively belonging to the faculty of Tune, be also in this state, and so with all the functions of every sense, organ, or faculty.

"In a natural state one or more of the functions or faculties may be, from some cause, rendered incapable of performing its or their proper functions, and, as it or they are more or less affected, we shall have the various phenomena, which are often exhibited from mere absence of mind, eccentricity, idiocy, etc., to perfect monomania and downright madness.

"Dr. Gall speaks of an organ of educability, or memory of facts, and another for the recollection of persons, etc.; while Dr. Spurzheim speaks, 'first of the faculties which perceive the existence and physical qualities of external objects, and those which procure notions of relations.'

"But this is an endless subject, and I will conclude this letter by recommending those who may hereafter engage in the above manner of investigating the faculties, to select the best subjects they can procure. Grown persons, if intelligent, should always be preferred, and the less they know of Phrenology the better." New subjects, who have had the proper explanations of the true nature of the state before they have entered it, should always be preferred to those who have been much experimented upon in the old way, as it is often difficult to remove habits or modes of thinking which have been acquired in this state; and it will, therefore, always be better to take those subjects who are best calculated in every respect, to give us the facts, unmixed with notions which have been previously acquired.

"The study, with all our advantages, *will be a difficult one*, and as we can only arrive at just conclusions by repeated experiments, I hope that those who may be engaged in the investigation will be guided more by a desire for the truth than an eagerness for renown."

Since writing the above letters, I have had many opportunities of witnessing similar results, and the experiments which I have since made, all go to prove the correctness of the remarks therein contained. I have made many experiments to ascertain the precise location of the organs, and my observations have generally gone to prove the correctness of most of the locations given them by Drs. Gall, Spurzheim, and other leading phrenologists.

I have witnessed a number of striking results in many of these experiments, and particularly so, when the subject had thrown the whole brain into this state, and I had requested him to awaken the portion of brain which phrenologists have denominated "*Self-esteem.*" If this is done properly, there is such an utter want of energy experienced on their part that it is almost impossible to get them even to raise an arm, and they have frequently declared that they felt as if it were out of their power to do anything. They seem to have lost all confidence in themselves, and did not feel as if they were any person, or could do anything of themselves.

Similar results, which corresponded to the faculties awakened, or put to sleep, were frequently witnessed; but the difficulty of obtaining subjects sufficiently interested in the matter themselves, has much retarded my progress. I shall, however, continue my experiments from time to time, and should anything occur, either to confirm or disprove this theory, I will state the one as freely as the other.

CHAPTER VII.

OF THE SENSES.

I CONSIDER the Senses to be fundamental faculties situated in the brain, each having a peculiar external apparatus or organization communicating with the external world—each of which is capable of receiving and transmitting sensations to its respective internal faculty. All these faculties, so situated, have respectively, power to attend to, perceive, judge, and remember, etc., their peculiar sensations—only as sensations. The respective qualities of these sensations must be attended to, perceived, etc., by the various other faculties, as the nature of the sensation relates to the faculty which can or has the power of perceiving it, etc.

I.—MOTION; OR, THE POWER TO MOVE.

This is also a distinct sense or faculty originating in the brain, and having an external apparatus—the muscular system—with, by, or through which the other organs produce motions peculiar to themselves. It, like all the other faculties, has functions peculiar to itself, and can perceive, judge, and remember motions simply as such—and as the muscular system is subservient to the will of all the other faculties—the force, direction, and continuation, etc., of the motions are regulated by motives in the various other faculties—and when an organ acts singly, the natural motion or language of the faculty acting is the result. Its being a distinct sense is very evident when it is in a somnambulic state. I shall, however, speak of it more fully hereafter.

CHAPTER VIII.

OF THE FUNCTIONS OF THE FACULTIES.

ALL the organs or fundamental faculties of the brain, I conceive, possess certain kinds of action, independent of what is called, more "intuitive perception," or knowledge obtained through the external senses, and that each faculty is composed of certain functions, which, together, constitute a faculty. The *peculiar* functions belonging to each faculty, properly so called, I conceive to consist of the following, viz.:—Consciousness, Attention, Memory, Association, Likes, Dislikes, Judgment, Imagination, and the Will.

These functions I conceive to be *peculiar* in each faculty, and that each faculty is only capable of attending to, perceiving, remembering, liking, judging, or associating, etc., those things or ideas which relate or are adapted to their peculiar capacities—and hold that it is impossible for any faculty to perceive, judge, or remember anything which belongs to the province of another. Benevolence cannot perceive size or form—nor causality or comparison, decay or the ridiculous. It is therefore plain that each must have its own *peculiar* attention, memory, judgment, etc., as well as its *peculiar* perception, etc., for it is well known that one organ may perceive and not remember at the same time that another does—and so with the other functions respectively. I shall, therefore, proceed to consider them individually, and first of:—

I.—CONSCIOUSNESS.

Consciousness is a knowledge of existence, and is the first act of the mind, and can be understood by the word *is*; viz.: to be conscious is to be sensible that something exists; but what that something is must be recognized by other functions. Consciousness simply acts, and it is a positive act; cannot be changed, and must remain simple in itself, and can only repeat itself; as, *a* is *a*, *is* is *is*, *I* is *I*, etc. *Is*, therefore, is the essence of mind, and must exist in all thought.

Mind identifies or makes two things the same; viz.; two a's are individually the same. True, *is* can only become itself, and is not limited by time or space, but in a sense creates both. A thing is itself, or, *a* is *a*; consequently the fundamental element of consciousness is necessary in all operations of mind where existence is manifest to the individual. But it is possible for all or any one of the functions in any of the organs of the brain to be active at the same time that consciousness is not (knowingly so); as we frequently see persons walking, speaking, or even singing, without being conscious of doing either. So, also, colors may be presented to the eye, scents to the smell, savors to the taste, and yet the person or persons may not be conscious in either case. But this is because the functions of attention and memory in the same organ or organs are not active at the time; and, without the action of these two functions in conjunction with consciousness, no action will be noted, and the individual will be conscious or not in exact proportion as they act together or not.

II.—ATTENTION.

Attention is that quality or kind of action in the mind which fixes it upon certain objects or things more or less intently as the function itself is active or not in the various faculties, and may vary in the different organs of the same brain, both as to strength and activity; but it does this without knowing what is attended to. It simply holds the mind to one or more things a sufficient length of time, however brief, to enable the other functions to act; and as it is exercised or not, our impressions are perfect or not.

Some persons are attentive in a very great degree to some things, and but moderately or not at all to others: and if this function is not active in a faculty, the organ cannot recognize anything.

III.—PERCEPTION.

Perception is that quality of mind which perceives a something, without knowing what is perceived; and as every thing in nature has certain qualities, it requires that those qualities should be recognized by the various functions of peculiar perception in those organs which have the power of perceiving them.

Take, as an example, a rose. Now as a rose has form, size, color, beauty, etc., these qualities must be perceived by the functions of peculiar perception in the organs of Form, Size, Color, etc., respectively, before we can know that it is a rose; or if we wish to remember, judge, or reflect upon it, compare it with others, etc., the functions of memory, judgment, etc., in the various organs, must remember, judge, and reflect upon those qualities which they severally can only recognize; and it is only after all these have acted, and have been associated the one with the other, that we can know that the thing presented is a veritable rose.

An idea may be produced internally by an act of the will, rendering the functions of memory or the imagination active, and thus produce an idea which may be recognized by the functions of perception in the faculty whose functions of memory or imagination have been active. As, however, different ideas may possess different qualities, they may make several impressions on the mind at one or nearly the same time. Ideas, therefore, may become complex under such circumstances, as the properties of the ideas double, or even abstract, as we may form an idea of a property or a quality independent of a particular idea itself. Both these kinds of ideas depend upon the activity of the functions belonging to the various other faculties which may be called into play; and, of course, may vary according to the nature of the previous ideas formed.

IV.—MEMORY.

Memory is that power which reproduces former cerebral impressions or perceptions that have been received and stamped upon its scroll, and is perfect or not as the impression or the attention at the time of reception was perfect or not.

It is notorious that some persons commit words to memory with the greatest facility, but cannot recollect persons, places, or events, etc., whilst others, remembering these, are deficient in committing words.

It is, however, sometimes very capable of cultivation, and if the attention be very active, it renders the impression more distinct, and, of course, the memory more perfect.

V.—ASSOCIATION.

This power produces a mutual association between the functions of the different faculties, and enables us to associate things with persons, localities, forms, numbers, colors, sounds, tastes, etc.; as a rose with a person, a person with a place, or a place with events, etc.

VI. AND VII.—LIKES AND DISLIKES.

The powers to like or dislike are as diversified as the faculties themselves. In individuality, they may like or dislike individuals; in eventuality, events; in color, colors; and in marvellousness, the wonderful.

Or if the faculty of individuality perceives a person —by the impression conveyed to that faculty through the sense of seeing—that person may be liked or not, according as these functions in the various faculties brought into play are pleased or not with the person's appearance, his qualities, or his behavior, etc. The function of dislike in the organ of size may not be pleased with his size, that of configuration with his form, or that of color with the shade of his hair; but the functions of love in the same faculties may be respectively so with the size of his head, the form of his mouth, or the color of his cheeks, etc.

VIII.—JUDGMENT.

Judgment is that act of the mind which decides upon the various impressions, actions, and qualities presented to the faculty to which it belongs, and in this is absolute; but when associated with the same functions in other organs, it constitutes relative or combined judgment.

"Judgment," says Dr. Spurzheim, "cannot be an attribute of every fundamental faculty of the mind, since the affective powers, being blind, neither recollect nor judge their actions.

"What judgments have physical love, pride, circumspection, and all the other feelings? They require to be enlightened by the understanding or intellectual faculties, and on this account it is that when left to themselves they occasion so many disorders. And not only does this remark apply to the inferior, but also to the superior affective powers: to hope and veneration, as well as to the love of approbation, and circumspection. We may fear things innocent or noxious, and venerate idols as well as the God of the true Christian. I conceive then that judgment is a mode of action of the intellectual faculties only, and not a mode of quantity, but of quality."

It is not necessary to take up the various propositions which are assumed in the argument preceding and following these paragraphs, as I think the Doctor has fairly admitted that these faculties have *judgment*, when he says that "they require to be enlightened by the understanding or the intellectual faculties." To "become enlightened," the faculties above alluded to must certainly be able to judge between things, qualities, and sensations, etc.; if they cannot do this, how are they to be enlightened?

Dr. Gall was of the opinion that every fundamental faculty possessed four degrees—or quantities—of activity: "the first was perception; the second, memory; the third, judgment, and the fourth imagination;" while Dr. Spurzheim believed the *intellectual faculties only* to be possessed of modes of action—not "modes of quantity," however, as Dr. Gall believed, but of quality.

I do not consider the action in the faculties, or in a single faculty, to be *degrees* or *quantities* of activity of the individual or whole organ itself, as Dr. Gall believed, nor yet with Dr. Spurzheim, that the same are modes of quality in an *undivided organ*, or that they are *confined* to the *intellectual faculties*. But that *all* the *faculties alike* are composed of certain *independent functions, each function constituting a part of the organ to which it belongs, and possessing a peculiar kind of action, which may differ in any of the organs, both in size, health, strength, and activity, or in quantity, quality, force, or energy, etc., according to circumstances.* I conceive that when an organ is active an emotion is experienced, and that degrees of activity are degrees simply of the *same* emotion. There are, therefore, as many peculiar emotions as there are faculties.

I therefore consider judgment a function belonging to every fundamental faculty of the brain, and its operation, like all the rest of the functions, is confined to the special functions of the faculty to which it belongs; and by an association with the functions of judgment in the other faculties, a judgment will result that is perfect in exact proportion to the soundness of the faculties acting.

IX.—IMAGINATION.

Imagination is that power which creates an image or embodies a thought, and therefore is entirely different from thought, which alone can conceive ideas, truths, and the infinite. You cannot imagine space, nor can you make a picture of it in your imagination. You can form an idea, or think of it, but you cannot imagine the infinite any more than you can measure it with a tape line; and although image-making and thinking are united when we think of things or impressions received through the senses, yet they are distinct and separate operations of mind. One is adapted to that which cannot be seen or touched, the other to things which are tangible. The imagination differs in the various organs in the same individual, and when a faculty is endowed in a high degree with this function, it is capable of originating new ideas according to the nature of the faculty to which it belongs; and when combined, or associated with the functions of the various other faculties, original plans, drafts, compositions, and machinery result. I consider constructiveness simply to be the faculty that adapts, constructs, builds, forms, puts together, either naturally or after a plan laid down or approved by certain combined faculties, and that it does not of itself invent machinery.

Some men do things only as they have been taught, others do them in a way of their own, untaught.

Birds usually build their nests in a certain form, without having been taught, and bees construct their cells of various sizes and depths, but their form, except the queen's, do not vary materially. It is true they have a natural plan or form of building their various cells, but they vary the shape, size, and depth of their combs according to the size and shape of the hive, or the space which they are to occupy. This cannot be the result of what is called instinct, nor the mere power to do, nor yet imitation, because they are never exactly alike. They must therefore invent or adapt the one to the other.

X.—WILL.

The will is that power which renders all the other functions active or passive, and is reciprocally affected by all the rest, particularly by the judgment; and when the will acts independently in all the faculties, contrary wills are the result.

To illustrate the operations of mind, let us suppose that an object or an idea of previous conception is presented. The function of consciousness being active, attention may be heedful, perception observe, memory note and store away, the likes love, the dislikes hate, the judgment distinguish, association unite or bring together, the imagination conceive, etc., as the will determines, influenced or not by the judgment, the peculiar likes or dislikes.

Now, if consciousness be active, the function of attention may act with the function of the will independent of any other function, but we cannot perceive without the action of the function of perception in unison with them. We may be conscious and attend, but until an object, a quality, or an idea is presented and noted by the proper functions capable of such recognition, we cannot know what we attend to; and it is only in proportion to the activity of consciousness and attention that we do this perfectly or not in any case.

We cannot see, hear, feel, taste, or smell knowingly, without the functions of consciousness, attention, and peculiar perception in these senses or organs are active; for it is notorious that we often pass friends, and are spoken to at times by others, without seeing the one or hearing the other.

When the functions of consciousness, attention, perception, and the will, have acted independent of the rest of the functions in the sense of seeing, this faculty has perceived a peculiar impression, and is conscious of the fact at the time; but if a person or a thing having peculiar qualities has been perceived the above functions in the organ of Individuality must also have acted. But before the peculiar qualities of light, or the size, form, and color of the person or thing can be known, these qualities will have to be perceived by the peculiar perceptive functions of the faculties individually, and are distinct ideas.

If, in addition to the functions of consciousness, attention, perception, and the will, the function of memory becomes active, the combination will not only be able to perceive and be conscious of the act at the time, but will be able to recall the idea impressed or stored away. We often see, hear, and learn things which we forget in process of time, because the attention, at the time of perceiving, etc., was not fixed, or the memory sufficiently exercised.

This is the reason why we may sometimes have a reminiscence, but not distinct memory. The function of memory in one organ may recollect and the other not; and thus we may know the name of an individual or thing, etc., but cannot utter it. In this case the memory in the organ of individuality remembers the person, the memory of eventuality remembers that the name was known, but the memory of language has forgotten it.

If, however, the function of association be brought into play or action with the rest, and by an association with the functions of the other faculties, the name which was lost by the memory of language may often be restored to that function.

An association or joined activity of the functions of one faculty with those of another produces an association of ideas; and we may associate a flower with a person, a person with a number, or a number with a place, etc., or an artificial sign may make us remember those which are absolute.

With the likes and dislikes added to these, the person or thing may be liked or not as the impressions upon these functions are agreeable or otherwise; and their pleasing or not is a distinct idea in either case.

It frequently happens that we dislike an individual at first sight; that is, his appearance may not please the function of dislike in one or more of the faculties; but upon a nearer acquaintance, his manners, or the qualities of his mind may act upon our functions of love in other faculties, and, by exciting them from time to time, we may lose our first impressions of dislike, and absolutely at last esteem the object of our previous hatred. The reverse is often the case with those whom we may at first love.

If the judgment be added to the rest or becomes active with them, it enables the faculties to judge of the person or thing; and the imagination may conceive improvements in either.

Correctness in all operations of mind will of course depend upon the health, size, and strength of the functions, and the amount of true knowledge previously stored away.

The will in the organ of motion and other faculties controls the muscular system, and when the faculty of motion is associated with any of the other faculties, peculiar motions can be produced; and as I have before stated, the natural language of the faculties as it is called, is the result of such combinations.

CHAPTER IX.

OF THE PECULIAR FUNCTIONS OF PERCEPTION IN THE DIFFERENT FACULTIES WHILE IN A NATURAL STATE.

ALTHOUGH this work is not strictly a treatise on mind, it seems necessary from what I have already said respecting the division of the faculties into functions, that, to complete my views of them, I should make some further explanations, or at least give my views of the *peculiar perceptions in the different* faculties, which perceptions respectively were heretofore considered to be the faculty itself; and, instead of being composed of separate and distinct functions, to possess as a whole only modes of action, viz.: modes of quantity or modes of quality.

I shall, therefore, here, to make the difference between the views of Dr. Spurzheim and myself more plain and easily understood, give his views numbered separately, followed by my own with corresponding numbers.

DR. SPURZHEIM'S VIEWS.

1. "I admit in the mind external senses, by which the mind and the external world are brought into communication and made mutually influential."

2. "The internal faculties are feelings and intellect."

3. "Both sorts may act by their internal power, or may be excited by appropriate impressions from without."

4. "The knowledge of our feelings is as positive as the experimental without."

5. "Every determinate action of any faculty depends on two conditions—the faculty and the object."

6. "The intellectual faculties are perceptive and reflective."

7. "The feelings and perceptive faculties are in relation and adapted to the external world, whilst the reflective faculties are applied to the feelings and experimental knowledge, and destined to bring all the particular feelings and notions into harmony."

THE AUTHOR'S VIEWS AND EXPLANATIONS.

1. I admit or consider the senses to be fundamental faculties situated in the brain; each having a peculiar apparatus or organization communicating with the external world, which are capable of receiving and transmitting sensations or impressions to their internal faculties. Which faculties, so situated, have respectively power to attend to, perceive, judge, and remember, etc., their *peculiar* sensations or impressions only as sensations or impressions. The respective qualities of these impressions, etc., must be attended to, perceived, etc., by the various other faculties as the nature of the impression relates to the faculty which can have or has the power of perceiving it, etc.

2. I admit that all the internal faculties possess like functions, which together constitute a faculty. But the particular function or power in each, whether it be called feeling or perception, I consider to be *peculiar in each;* viz.: the faculties called the "affective faculties or feelings" by Dr. Spurzheim, and by him divided into "feelings proper to man and animals," and " feelings proper to man;" the essential nature of which, he says, is "only to feel emotions," I consider to be like all the rest, whether termed intellectual, reflective, or otherwise, possessed of peculiar powers to observe, perceive, know, or recognize, etc., the *peculiar* impressions, sensations, objects, or ideas, etc., which their individual capacities render them capable of. Whether in Destructiveness, this power be called the power to observe, know, or recognize destruction or desolation; in the organ of Benevolence, generosity; in Conscientiousness, justice; or in Causality, the cause of either.

3. I consider that all the perceptions, etc., may be excited by internal as well as external impressions.

4. I consider that the knowledge of our internal impressions, whether they are called feelings, perceptions, or ideas, etc., are as positive as those from without.

5. I consider that every determinate action of any function depends upon the function and the impression.

6. I consider that all the faculties perceive and reflect, or judge those things which relate to their capacities.

7. I conceive that all the faculties are in relation and adapted to the external world, and all may receive impressions from within or without, and each may reflect, etc., upon its own peculiar impressions, which is accomplished by the other functions belonging to each faculty. The power to feel, perceive, or know impressions in the organ of Destructiveness, may perceive the impressions relating to desolation, decay, or destruction.

The same function in the organ of Combativeness may perceive the impression relating to resistance, quarrels, battles, contests, etc.

The same in Benevolence may recognize the benign, the kind, the generous, or liberal.

The same in Reverence, what is venerable.

The same in Individuality may perceive persons or things.

In Eventuality, events.

In Tone, tones.

In Comparison, the identity or difference between them; and in Causality, the cause of one or of all.

If the function of perception in the faculty of Destructiveness note or observe an impression, whether of desolation, decay, or death, its peculiar functions of consciousness, attention, perception, and the will, must have acted; but before it can judge of the impression or remember the same, its functions of judgment and memory must also act; and so with the likes and dislikes, its imagination, and its association before it can like or dislike the impression, imagine another, or associate it with any of the other faculties, which would enable the mind to draw conclusions as to the exact kind of desolation, decay, or death; its identity with other desolations, etc., or the cause of either.

If the function of perception in the organ of Causality perceives a cause, whether it be from an external impression or an idea of the mind, the same functions in the organ of Causality must have acted before it could have been observed, judged, remembered, liked or not, another imagined, the one or both associated and compared with others.

It therefore seems very evident that if an impression is received through the external apparatus, or organization of the sight, by the function of perception in the sense or faculty of seeing, that impression may be judged, etc., by the other functions of that faculty only as an impression, and before the respective qualities of that impression can be known; the functions in the organs capable of recognizing them must also attend to, perceive, judge, and reflect upon them.

Thus: if the impression received be caused by the presentation of a rose, the impression conveyed at the time will be perceived, judged, etc., by the faculty of seeing; individuality will perceive, judge, and remember the object's distinct or individual existence; size, its bulk; color, its shade; form, its configuration, etc., before the mind can know that it is a rose.

If it be beautiful, that quality must be recognized

by Ideality. If decayed, by Destructiveness; its identity with others by Comparison; and the cause of either, by Causality.

As each function, however, may differ in health, strength, and activity, or in quality, quantity, and energy, etc., it must follow that there will be degrees in their capacities.

Therefore if the functions, particularly that of love, in the organs of Combativeness and Destructiveness, be large and very much excited by impressions received or impulses given by the other faculties, the consequence will be violent emotions in these faculties, which, if not held in check by other functions in counteracting faculties, injuries will be contemplated, and if aided by the power to do, injury may be done purely for the love of it. If with the above faculties the functions in the organ of Acquisitiveness be very active, the motives to do injury would be for gain; if the same in the organs of Cautiousness and Secretiveness be added, the disposition would be to do the deed as an assassin, and to cloak or hide it when done.

Should the functions of the imagination in all the faculties be large and active, the plan to do the same would be likely to be good or well-contrived; if not, the reverse would be the case, and so with the rest of the functions respectively.

The analysis and synthesis might be carried on at pleasure, but there has necessarily already been considerable repetition, which I cannot yet wholly avoid, as I have still to consider the functions while in a somnambulic state.

I.—OF THE PECULIAR FUNCTIONS OF PERCEPTION WHEN IN A STATE OF ARTIFICIAL SOMNAMBULISM.

When a function of perception in any of the faculties belonging to the brain becomes active while in a state of Artificial Somnambulism, it is enabled to perceive without the aid of the external senses, and the *perception thus accomplished* I call *Clearmindedness.*

I conceive that this power is possessed by *all* the faculties while in this state; and that they can individually only perceive what relates to their peculiar capacities, whether it be an idea in the mind of another, or is composed of matter and exists in the external world, if their individual attentions, etc., be directed to them.

The functions can and often do act independent of one another, and as they act singly or not the external or visible signs or results, etc., differ also: viz.: if the function of perception in the organs of Motion and Imitation become active together, independent of the functions of consciousness and the memory in either of these faculties, an imitative motion may take place or be produced *independent of the subject's knowledge;* as is frequently the case, although the rest of the faculties may not be clairvoyant, when drawing the attention of these faculties to a motion which we wish to be imitated.

This can sometimes be effected without saying a

word, particularly if a slight noise be made by that motion, which, through the hearing, attracts the attention of these faculties. This, however, cannot be effected unless the attention be first drawn by some means; but if the attention in these faculties be watchful or attending to the external world, etc., it is astonishing how slight a hint *or even a thought* will be observed, and produce the desired results without the subject's knowledge.

This peculiar power of the faculties while in this state, enables subjects, or persons who are *supposed to be "impressible"* and under the influence of a nervous fluid, to know or arrive at correct conclusions respecting the contents of sealed letters, etc., or even to arrive at a knowledge of the thoughts of another, although the person thinking or the letter to be read be at a distance.

I could furnish innumerable facts to prove their ability to do these things were it necessary. I will conclude this section by relating a case in point which lately came to my knowledge:

A lady who, after having been in a somnambulic state, awoke with the impression that she *must* or could not avoid knowing the mind of the operator. Since that time she has been possessed of that power, and her mind, which she thinks *must be*, is, of course, almost always directed to him; consequently, when so directed, her mind enters the somnambulic state involuntarily, and she is, as it were, irresistibly compelled to know his thoughts and his whereabouts, etc.

This power, which is a great annoyance to her, will be likely to continue as long as her belief remains the same, or until she learns the true nature of her state, and that the remedy lies within herself; viz.: to prevent her mind from entering the state, or of reaching out, *clairvoyantly*, after him.

I have a number of subjects who can do the same thing at *pleasure*, independent of any one; and if the above lady had had the proper instructions before she entered the state, the power which now annoys her she could have used at pleasure.

II.—THE FUNCTIONS CONSIDERED WHEN IN A STATE OF ARTIFICIAL SOMNAMBULISM.

1.—*Consciousness.*

Consciousness and sensation are completely under the control of the will in most subjects while in this state, and are extremely active or entirely passive, as the will of the subject determines.

2.—*Attention.*

When persons are in the sleeping condition of Artificial Somnambulism, all the senses and faculties lie dormant or inactive, and it requires an express action of their will to render any of them active. They can do this whenever they please, either partially or entirely, but they cannot see, and hear, and smell, and feel, etc., at one and the same time. But they can see, or not, hear, feel, taste, smell, move, think, attend, perceive, be conscious, remember, judge, imagine, like or not, etc., as they please, or when they please, independent of any one.

In the waking condition of this state, their attention is commonly directed to the person into whose care they have entrusted themselves, not because they cannot do otherwise, but because they choose to do so, and often do not wish to be disturbed by others; and as it commonly requires an effort for them to do anything (particularly when they have entered it for the first time), it is necessary when an experiment is desired, that their attention should first be drawn and properly directed and their full consent to do so obtained before the experiment is attempted. I expressly again deny that there is anything like what is called sympathy between the subject and the so-called operator; and *insist* that they never do anything by sympathy, nor can they perform any experiment if their attention be not first directed to the object, either by the instructor or the person into whose charge they have entrusted themselves, or by their own will. They *cannot* and *do not taste* what is in the operator's mouth *unless their attention be first directed to his tasting*, etc.; and when that is done, it is *not* by sympathy but by *throwing their mind there* and tasting it. They can do the same thing if the article to be tasted by them be in the mouth of *any other person*, or is placed in *a box at a distance*. Their information, therefore, cannot be obtained by sympathy, unless it be admitted that they can sympathize with a box or any other inanimate substance.

3.—*Perception.*

The powers of perception in this state compared with the same function in a natural state are inconceivably greater, and it is impossible for those who have not seen or made the necessary experiments to conceive the difference. Language fails to express it, and our common philosophy is too circumscribed to explain the reality.

This function, when aroused and properly directed, is extremely sensitive and correct, and most subjects by an act of their own will can translate their perceptions, etc., to any part of the body, whether to the stomach, feet, hands, or fingers, and use them at these points as well as at any other. The same thing is often witnessed in cases of natural somnambulism, and only exists because the somnambulist's mind has from some cause been directed to these parts.

They can also translate their faculties to a distance, and I have had them to perform thousands of experiments correctly at various distances, varying from ten feet to eighty miles, independent of any previous knowledge or communication whatever, either personal or otherwise. I state these facts in the face of all the learning, opposition, prejudice, disbelief, and ridicule of the age; and would ask those unbelievers, who are satisfied with the philosophy which cannot even tell them what light is, to say where their powers of perception shall cease, when it has been proved that these things have and can be done at a distance of over and above sixty miles.

Their powers of perception, however, are not always infallible; that is, subjects do not always tell correctly. Their not seeing, tasting, smelling, etc., truly sometimes, is owing to their own imagination; because when persons are in a state of Artificial Somnambulism, they can see, taste, smell, etc., what they imagine, as well as they can that which really exists; and, therefore, if they are not very careful to look before they imagine, they may see or taste falsely respecting what exists, but yet truly what they imagine. It is very difficult to tell when they do the one and when they do the other, and it is yet to be learned whether cultivation will produce perfection. Practice will no doubt much improve it, and I have always observed that when the subjects were themselves interested in looking or tasting, etc., the result was more satisfactory; showing that it requires that they should not only guard against their imagination, but that it also requires their whole attention to perceive correctly. If they are indifferent or unwilling to perform experiments, their answers cannot be depended upon.

I hold, therefore, that if subjects in this state can use their senses and faculties correctly at a distance of sixty or eighty miles, that we cannot limit their abilities short of anything which would be incompatible with this philosophy.

I wish, however, to be distinctly understood upon this point, and contend that if they can tell things which are placed or are passing at a distance of sixty miles correctly, they can do so at any distance under like circumstances. *But I still, knowing that they may and often do imagine, would not receive their evidence as positive proof; nor would I absolutely declare, in any instance, that they did not see correctly unless I could prove the contrary, knowing also that they have done these things correctly, and can do so again if they direct their minds properly.*

Practice and the proper cultivation of this faculty will I am persuaded, at no very distant day, so improve this power, that it will not only rival the powers of the natural eye, but so far exceed them that a comparison between them will be in favor of clairvoyance. This may at this time appear a sweeping assertion, but from what I have seen I am constrained to predict that, instead of being limited to its immediate surroundings, like the natural eye it will be able to peer through matter, darkness, and space, until there shall not be a nook in all creation that its power cannot reach or its observation scan.

The clearmindedness of the other senses is also susceptible of the same improvement.

4.—*Memory.*

The activity of the memory in this state depends entirely upon the will of the subject, and in passing from this state into the natural by an act of their own will, some can entirely forget what has taken place in that state, or remember only as much as they please.

If, when a subject is in this state, nothing be said about their remembering, or they do not make an express effort or desire to do so of their own accord, they remember nothing when they awake.

I have in several instances forgotten to request persons to remember, and when they awoke it was impossible to make them believe that they had been asleep at all, or had done anything. This is particularly the case when they have entered the state for the first time, and have not yet learned the necessity of doing so.

One subject in particular, whom I had forgotten to remind, had taken a walk in the garden, played upon the piano, sang, danced, and named many individuals placed behind her chair, etc., yet could not be made to believe it until she had again entered the state and was told to remember.

Another lady, who had also done many similar things while asleep, could not, after she awoke, be persuaded that she had been asleep at all, or had done anything. She believed she had been imposed upon, although her friends present all testified to the contrary. She has never since consented to enter the state again, and still believes that she never has been in it. This case shows the necessity of getting them to remember before they awake; and it should always be done when they can be persuaded to do so, but I have had some who positively refused, and of course knew nothing when they awoke.

5.—*Association.*

Association, like all the rest of the functions, is controlled by the will, and is active or passive as that function is active or not.

6 and 7.—*Likes and Dislikes.*

The likes and dislikes in this state are easily excited and are respectively easily pleased and displeased; and when anything occurs to please, the subjects are generally courteous and affable, but *always thoroughly independent.*

On the contrary, when anything occurs to displease, they can resent, despise, or be indignant in the extreme, as the nature of the case may be.

8.—*Judgment.*

The judgment in this state is correct or not according as the thing to be judged interests them or not. When active it is extremely correct, and the reverse when the opposite is the case.

9.—*Imagination.*

The imagination, when unrestrained by the will in this state, is extremely active; and as they can see what they imagine as well as they can perceive what really exists, as I have before stated, it is difficult to know when they do the one or the other; or whether they do not do both at one or nearly the same time.

The following is a case of the latter description:

I requested a lady, at the suggestion of a certain gentleman, to visit his house for the purpose of testing her powers of clairvoyance, neither of us knowing where he intended to be, or who were to be there, etc. She complied willingly, and said that he, the gentleman who made the request, was in the front room in company with two other gentlemen, naming both, and that he was speaking to Mr. A.

We *at once* sent to ascertain the facts, and found that the owner of the house was not in *that* room, but that the other two gentlemen whom she had named *were*, and speaking together.

In this case, she being desirous of finding a certain individual, her mind was placed upon him and imagined that she saw him in the front room, but absolutely did see who really was there. Had she been upon her guard, and looked without imagining, she would have seen that the owner of the house was not there, and would not have been liable to the oversight she made by imagining first and looking afterwards.

I here again repeat that when persons are *ignorant* of their powers while in this state, their imagination is easily imposed upon, and they can be made to imagine and see what they imagine by operators as ignorant as themselves; not because they must do so, but because they *believe* that they must. The same things have taken place frequently in excitable subjects when awake, independent of any operator, and it is therefore not extraordinary that they should occur when all the faculties are excited by an operator who has imposed upon their credulity. But I contend that they are not and cannot be produced awake or asleep, contrary to the subject's will, and that when they do take place they are the effects of an excited imagination simply, and not real perceptions of what actually exists. The same impositions are often practised upon their hearing, feeling, taste, and smell, etc., by some operators, who *will* ice to be hot, water to be brandy, hartshorn, or cologne, etc. I have no doubt the subjects do hear, feel, taste, and smell what they say, and believe the one to be the other, but it is upon the principle that they see what they imagine. But their believing or doing so does not prove that it is effected by the will of the operator, any more than their believing so should make the water brandy, etc. The effect is produced by reading the mind of the operator and blindly consenting to be governed by it: and they could just as well imagine the one to be the other independent of him or contrary to his will, if they were so disposed.

Such impositions are therefore not only ridiculous and useless, but imprudent, and I have no doubt that these and similar inconsistencies have disgusted and turned many fine minds from this interesting science. I hope, however, that time and a thorough exposition of such absurdities will reclaim their favor, and make the science worthy their attention.

10.— *Will.*

The will is paramount in this state, and controls the activity of all the functions.

The will to see, hear, feel, taste, smell, or move, etc., depends upon the activity of this function, influenced or not by the judgment, in one or more of the faculties, and is exercised or not, according to the determination formed to do the one or the other.

It is supremely independent of the will of the operator, or any other person; and when the subject is acquainted with the true nature of the sleep and his powers therein before he enters it, it is impossible to impose upon or dally with him—much less while in this state than when awake.

I therefore positively deny that it is possible for any person to do anything with them in any way contrary to their will, or that they would be more likely to yield to arguments or persuasions in this state that when awake. I have always found the reverse to be the case, and have generally had much difficulty in getting them to perform experiments, especially if they have had a proper explanation of the nature of the state before they enter it.

CHAPTER X.

OF READING OR KNOWING THE MIND.

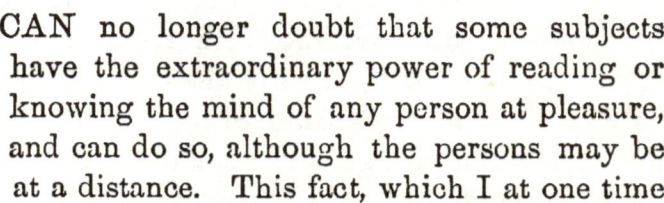CAN no longer doubt that some subjects have the extraordinary power of reading or knowing the mind of any person at pleasure, and can do so, although the persons may be at a distance. This fact, which I at one time very much doubted, I was eventually forced to believe, and the only rational solution of the manner in which it is effected is, that the mind, or the faculties of the brain reaching out, or being translated to the mental aura of the individual, there reads, and it may be often, unconsciously, the mental image formed, either in symbols or in words. That they can do this, I am constrained to say, I have proved more than a thousand times. I will give a few illustrations in detail.

I.—ILLUSTRATION.

When on a visit to S———, the following instance of this power was exhibited to a private company at that place:

Subject, Miss W———. First sitting. She was over an hour in entering the state, and after I had declared her to be in a perfect state of Artificial Somnambulism, I was called out of the room by a reverend gentleman of the place, who requested me, in a very opinionative and sarcastic manner, to will Miss W—— to descend into the parlor below, and bring up a tumbler of water which he had placed upon the table—for the purpose of testing her powers of reading the mind—and insisted that it must be done without uttering a word.

As I had doubts of succeeding, this being the lady's first sitting, I stated very frankly to the gentleman that I would try the experiment to gratify him, but that the success of it depended very much upon, whether she was paying attention to my will or not, and that I almost always found it necessary to draw their attention before I attempted any experiment.

As, however, nothing would satisfy him but the plan he had laid down, I told him that if he would open the doors between us, so that she could pass down-stairs, I would attempt it from where I stood.

After having done so our relative position was as follows:—The chair upon which she was seated was in the middle of the next room, with an entry and stairway between us, and her back towards me. I willed her first to get up, and then successively to pass down-stairs, take hold of the tumbler and bring it up.

This, to my utter astonishment, she did to the very letter, and when she had brought it up to me, speaking aloud, I requested her to give it to a certain gentleman, without naming him. She followed the reverend gentleman through the crowd and around the room several times, until she obliged him to take it. During the flight of the reverend gentleman many others attempted to deceive her, by reaching for the tumbler, but she would give it to none but him.

II.—ILLUSTRATION.

Before I enter into the details of this case, it will be necessary to make some remarks upon the practice of putting persons into what has been called *en rapport*, communication, correspondence, or sympathy with them, etc. Various methods have been adopted to effect this end, and most of my readers are, no doubt, acquainted with them, I will not enter upon a description here. I have always found such proceedings unnecessary, and when such a desideratum is desired by any person, it is only necessary for the instructor, or the person into whose care they have entrusted themselves, to direct their attention to the individual in question, or to make them acquainted with their desires, and they will travel with, or read their minds as well *without contact, passes, etc.*, as with them.

The reason why the interference of the operator, as he is called, is sometimes necessary, is because the subject has entered this state under his care, and commonly does not attend or listen to any but him, and, therefore, does not hear, unless directed by him to the person speaking, or they do so of their own accord. If, when a person is in this state, he brings back, or places his mind upon the room in which he is seated, he can hear what is said as well as if he were awake. The supposition or belief that he cannot hear is a mistake, and in performing experiments it is necessary to be extremely cautious not to speak of what they are desired to look at, for, if their attention is directed to the room, they can hear the slightest whisper, although it be spoken at the most remote part of the room. I shall now proceed to give the second illustration in detail, and extract the following from notes taken at the time. Subject, Miss Z——, and her tenth sitting. She was requested to travel with Mr. F., and having consented, was asked by him:

Mr. F.—"Where are we now?"

Miss Z.—"On a large water, in a steamboat," and pointing up, said, "There is a flag, it is striped."

Mr. F.—"Well, where are we now?"

Miss Z.—"In the cabin. It is very beautiful."

Mr. F.—"Where now?"

Miss Z.—"Looking at the machinery." She gave a description of various things about it, and, without being asked, said, "We are on deck now."

Mr. F.—"Well, what am I looking at?"

Miss Z.—(Seemingly much delighted)—"I see another boat coming towards us, but it looks very small. It is very far off, and creeps along like a turtle."

Mr. F.—"Where are we now?"

Miss Z.—"If you cannot take me to a better house than this, I will not go with you any more."

Mr. F.—" Why, what kind of a house is it?"

Miss Z.—"It is a poor concern, and is made of rough boards. Let us go."

Mr. F.—" Where are we now?"

Miss Z.—(Very much pleased and laughing heartily)—" Oh! just look at them. See how busy they are."

Mr. F.—" What do you see?"

Miss Z.—" Why beavers, to be sure. Look how they are building their huts."

Mr. F.—" Where are we now?"

Miss Z.—" I think in a city. We are before a large house that has a sign; see how it swings."

Mr. F.—"Let us go in."

Miss Z.—" No, I would rather be excused. Let us go home."

Mr. F.—" Well, where are we now?"

Miss Z.—" This is a strange-looking boat, but now we are on a better one. There—now walk out on that plank. Here we are at home again."

Mr. F—— then stated that she had read his mind correctly; and that he first imagined himself in a steamboat upon a lake, he then threw his mind into the cabin, then upon the machinery, then on deck, and imagined a steamboat in the distance. He next imagined himself upon an island, where the boat usually stopped to take in wood, and upon which there was a shanty, which she described. He next threw his mind upon beavers at work, constructing their huts. Then in Millwaukie, before the hotel, and finally, upon an inferior boat used to convey passengers to the main boat, and then home.

That they can read the mind, or see and know what is passing in that of another, is a question which has been much agitated, but never, I believe doubted by those who have made the proper experiments.

It is astonishing with what facility some subjects follow, or read the minds even of strangers who may desire to take them to places where they have never been; and when there with what accuracy they describe places, persons, or things existing or passing at the time.

The following is an illustration of this kind of seeing:

Mr E—— was desired, at the request of a gentleman, to visit his home with him—which was distant about fifty miles—and when he had followed him by reading his mind, he described the peculiarities of the mill and the *house attached to it*, the number of rooms in the house, where entered, the furniture and relative position of the same, his wife, whom he described as being slim, tall, with very dark hair and dark complexion, dressed in a brown gown, having a child in her arms. Another child, of about four years old, was described as running about the room; and an old gentleman, rather portly, bald, and dressed in drab clothes, was seated upon a settee. All this the gentleman declared was correct, and could not have been better described by the subject if he had been there in body at the time; and, as the gentleman had never seen the subject before, nor the subject either him or any of the family, he was convinced, though skeptical before, that he must have just seen what he described.

The description of the above residence and family was so minute, so clear, and so unhesitatingly done, that if it, or a like description, had been given to the most skeptical, it must have convinced him that there was something more in their powers than "is dreamed of in the world's philosophy."

ANOTHER ILLUSTRATION.

Subjects, Miss Z—— and Miss S——, both in the state at the same time. They were requested, by a lady present, to visit her home in company (eighty-two miles distant), and after obtaining their consent—as neither of the subjects had ever been there—she conveyed them there in thought, and desired them to name the number of trees before a certain house she was looking at.

Answer.—"Four."

Ques.—"What kind of a house is it?"

Ans.—"A two story brick."

Ques.—"How many steps are there at the front door?"

Ans.—"Only one."

Ques.—"How many at the back door?"

Ans—"Three."

Ques.—"How many rooms are there on the first floor?"

Ans.—" Four, besides the kitchen."

They then described the furniture in the various rooms, the mantle ornaments, the carpets, and which rooms were papered, and which were not, etc.

Ques.—" How many rooms are on the second floor ?"

Ans.—" Five."

They described one of the rooms very minutely, stating that it looked more like a kitchen than a sleeping or sitting room, and that it had a strange shape, and the pitch of the roof over it was very peculiar (many things belonging to a kitchen were in it at this time). They named the number of beds and the kind of furniture in the other rooms, and when taken to the yard, described the relative position of the trees, the flower-beds, and the peculiar make of fence, etc. All of which was admitted to be correct.

THE FOLLOWING IS ANOTHER ILLUSTRATION.

Subject—Miss K. Z.

She was requested by the Rev. Mr. I—— to take a journey with him, and having consented, the following questions were asked and answered :

Ques.—" Can you tell where we are now ?"

Ans.—" In the cars."

Ques.—" Where are we now ?"

Ans.—" Somewhere down the country. I have never been here before."

She then, at his request, described the house, its situation, and its peculiarities, etc.

Ques.—" Who do you see in the house ?"

Ans.—" I see but two in the back room."

Ques.—" Will you describe them ?"

Ans.—" The one is a lady. She is small, has dark hair and eyes, *and has lost four of her front teeth.*"

Ques.—"Correct. And the other ?"

Ans.—"Is a gentleman. He is stout, middle-aged, wears glasses, and is now engaged in reading a newspaper."

I then asked Mr. I—— whether he was sure that she had never seen the persons she had described?

He stated that he was satisfied that she had not. I then requested him to think of the first name of the gentleman, and she would be able to tell him what it was by reading his mind.

This he seemed persuaded she could not do, and stated that if she did so, he would be obliged to believe in the science. Having stated that he was thinking of it I requested her to state what name he was thinking about. She stated that the name was Charles W——.

Rev. Mr. I——. " Yes. And the lady's ?"

Ans.—" Mary ——." He was perfectly satisfied.

I could illustrate this peculiar power by a number of other cases, but consider it unnecessary, and will proceed to examine the theory of Dr. Collyer respecting this phenomenon.

THEORY OF DR. COLLYER.

With respect to the theory of reading the mind at a certain angle, I have but to say that it does not accord with my experience. As some persons, however, may not have seen this theory, I quote the doctor's own words, printed in the *"Albany Argus,"* as follows:—

"I have always advocated the philosophy that the "nervous fluid was governed by the same code of "laws which governed light, heat, etc., as radiation "and reflection, and actually made a lady perform "the same class of phenomena which is the wonder "of travelers in the east. She was desired to look into "a cup of molasses (any other dark fluid will answer "the same purpose), and when the angle of incidence "from my brain was equal to the angle of reflection "from her brain, she distinctly saw the *image* of my "thoughts at the point of coincidence, and gave min-"ute descriptions of many persons whom she had no "idea of. She saw the persons and things in the "fluid only when the angles of thought converged." Again: In the doctor's work on "Psychography," etc., page 39, he remarks:—"I was not a little pleased on discovering that in Europe, some months after my 'bowl of molasses' experiment, the subject of PSYCHOGRAPHY was occupying the attention of the ablest minds. The public will, therefore, suspend judgment, more particularly as I have experimentally shown the fact of the *embodiment of thought* before audiences of several hundreds in the city of Philadelphia, with entire success.

"Only a few nights since it was repeated. The gentleman who came on the stage depicted on his mental vision, 'a horse, a man and a horse,' which the recipient (Evan) gave loud and distinctly to the audience. Another gentleman came forward—the recipient declared he saw 'a large white marble building.' The United States Bank was the object thought of. On another, Mr. E—— came on the stage, and directed his thoughts into the 'bowl of molasses.' The recipient declared he saw a tall, stout, elderly gentleman with white beard and gown on; he then saw a marble statue. The correspondent declared to the audience, that he was thinking intently of Lawyer Chancey, who corresponds exactly with that description; and in the second instance, he was thinking of the marble statue of one of the lions, at the Exchange.

"These results have put the matter beyond all cavil and doubt; I can repeat them any time under favorable circumstances, such as are laid down in the pages of this work."

Doctor Collyer has, no doubt, done much towards the advancement of the science in many parts of this country, and I am sorry that he has, by drawing hasty conclusions, left himself open to the shafts of those who are disposed to ridicule it.

For my own part, I never doubted that his subjects saw what he has described, nor do I question the veracity of those concerned; but *if* the doctor had made the necessary experiments he would have soon found that the same could have been done *at any angle* without the aid of "that bowl of molasses" or any other agent, and that his subjects could have done it as well by looking directly at his brain or mind as by any other method. The reason why his subjects could not (apparently) do it at any other angle than the one he stated is, because he pitched upon that angle, and they *believed* with him, that it could not be done at any other. Whether he made them *believe* so, or it was a *belief* originating with themselves, I will not pretend to say. *But certain it is* that, if subjects *believe*, we are made to *believe* that a certain thing cannot be done (either before or after they enter this state) they cannot do it for the simple reason that they *do not try*, and have predetermined that it is impossible.

These are the facts, and it is very plain, that their looking and seeing at that angle only proves that they threw their minds there, and *believing* that they could see, did so. If, however, they had known the facts and their powers they could have thrown it to any other place, and have seen just as well under like circumstances.

Their being able, also, to see, hear, smell, and taste, etc., from other parts of the body than which are used for those purposes in a natural state, will be the subject for another chapter.

MENTAL ALCHEMY OR ELECTRIFYING.

As there are still exhibitions in various parts of the country of the condition which it has pleased some to denominate "Mental Alchemy," and which is supposed to be induced by electricity, but is, in fact, only an offspring of the sympathetic doctrine, it may be proper here to say something in regard to it, which, however, will only be a repetition of what I have before stated. "Mental Alchemy" is simply a partial state of artificial somnambulism, in which some of the functions in certain organs act independent of the judgment and the will in the same faculties; consequently, the subjects become credulous in the extreme, and not knowing that they can do otherwise are made to *believe* that white is black; water, brandy; ice, hot; and other ridiculous contraries that the operator may imagine, or others invent for him.

This, to say the least of it, is turning a useful science into ridicule, and is, at best, but folly in the extreme. It is a matter of regret and surprise to me to see those who make pretentions to intelligence, night after night encouraging, by their presence, scenes which are too ridiculous to draw anything from sensible persons but a smile of pity, as the subject is not himself; and even a Daniel Webster, under like circumstances would, for the time being, become idiotic. From what has been said in previous chapters, it is very evident that electricity can have no agency in producing this condition, and it is equally absurd to suppose that the so-called operator can have absolute power over any one in this state, if the subject be properly instructed, or has a true knowledge of the facts. The apparent power exhibited is the result of a blind belief and non-resistance on the part of the subject, which a proper understanding of the case would dissipate to the four winds. In arousing them from this condition it is only necessary to draw their attention to the fact. Any exclamation on your part such as, Arouse! awake! or "All right!" will bring them to themselves again. This they could also do themselves if they would make the effort, and it is only because they have been taught otherwise that they permit themselves to be directed. But apart from the ridiculous position in which a subject is placed, much injury may be done to the brain by inexperienced persons, and the public would do well hereafter, to be cautious and reflect upon the consequences which might ensue, viz.:—*that of remaining an idiot*, before they trust so delicate an organ as the brain to the tender mercies of those who choose to play with it regardless of consequences, so that their own aggrandizement has been effected, or money is put into their pockets.

CHAPTER XI.

I.—OF THE IDENTITY OF OTHER MYSTERIES WITH THIS STATE.

MANY things are ascribed to a supernatural or mysterious power, even at the present day, which are identical with this state; and among the greatest of antiquity is the Oracle of Apollo of Delphi.

It is not necessary for me to give a history of this oracle here, as I presume most of my readers are already acquainted with it, and it will be sufficient for my purpose to note that which has a direct bearing upon the question. I quote the following account of the manner in which a priestess was affected, after having been placed upon the tripod, to show the identity of her state with that of a person in a state of Artificial Somnambulism:

"Great preparations were made for giving mysteriousness to the oracle, and for commanding the respect paid to it. Among other circumstances relating to the sacrifices that were offered, we may observe that the priestess herself fasted three days, and before she ascended the tripod, she bathed herself in the fountain of Castalia. She drank water from that fountain, and chewed laurel leaves gathered near it. She was led into the sanctuary by the priests, who placed her upon the tripod. As soon as she began to be agitated by the divine exhalation, her hair stood on end, her aspect became wild and ghastly, her mouth began to foam, and her whole body was suddenly seized with violent tremblings. In this condition she attempted to escape from the prophets, who detained her by force, while her shrieks and howlings made the whole temple resound, and filled the bystanders with sacred horror. At length, unable to resist the impulse of the god, she surrendered herself to him, and at certain intervals uttered from the bottom of her stomach, some unconnected words, which the prophets ranged in order, and put in form of verse, giving them connection, which they had not when they were delivered by the priestess.

"The oracle being pronounced, she was taken off the tripod, and conducted back to her cell, where she continued several days to recover herself from her conflict."

The priestess in the above account (with the exception of her agitation and violent tremblings) was not exactly affected like those who enter this state at the present day; but was identical with those who entered it under the care of Mesmer and his immediate followers.

Such effects were in *those times* considered necessary particularly so in the case of the priestess, where it was of the greatest importance to give a mysterious character to the oracle, for the purpose of commanding respect and of filling "the bystanders with sacred horror."

Mesmer, considered it indispensable for the cure of diseases, and the subjects in both cases were made to *believe*, that certain effects were to follow the respective operations, which of course took place.

I have frequently seen persons similarly affected, who had erroneous notions of the state, and I could produce like effects upon ignorant persons at almost any time; but it is highly improper and might be injurious.

I have always found that the more intelligent the subject is, and the better the nature of the sleep is understood, the more he is at ease, and the less is seen of these unpleasant and unnecessary symptoms.

With respect to the nature of the vapour or gas, which is said to have issued out of the mouth of the cavern: I have nothing to say further than, that if there was any escaping—that it was not in, my opinion, the nature of the gas which produced these effects; because we know of none now that would; those of the nitrous oxide being quite different. Coretus, who, it is said, first discovered its effects upon goats, prompted by curiosity, also approached the mouth of the cavern, and found himself seized with a like fit of madness, "skipping, dancing, and foretelling things to come." But, as we have no evidence upon which we can depend, this skipping, etc., being natural to goats, and not agreeing with its effects upon the priestess, we may think of it as we would of all other things that have been said about it, of which the following is a specimen:

"This place," speaking of the mouth of the cavern, "was treated with singular veneration, and was soon covered with a kind of chapel, which Pausanius tells us was originally made of laurel boughs, and resembled a large hut. This, says the Phocian tradition, was surrounded by one of wax, and raised by the bees." If we can believe this, we may believe all that has been said about it.

II.—OF THE MYSTERY PRACTISED BY THE MODERN MAGICIANS OF EGYPT.

The mystery practised by the modern magicians of Egypt I ascribe to the somnambulic condition, and, although it is shrouded by magical incantations, charms, spells, and a host of other unnecessary accompaniments, it is easily sifted and brought within the bounds of reason and philosophy.

The following account, taken from Lane's work on Egypt, will give the reader an idea of these magical experiments:

"A few weeks after my second arrival in Egypt, my neighbor, Osman, interpreter of the British Consulate, brought him to me, and I fixed a day for his visiting me to give me a proof of his skill, for which he is so much famed.

"He came at the time appointed, about two hours before noon, but seemed uneasy, frequently looked up at the sky through the window, and remarked that the weather was unpropitious: it was dull and cloudy, and the wind was boisterous. The experiment was performed with two boys, one after another. With the first it was partly successful, but with the other, it completely failed. The magician said he could do no more that day, and that he would come in the evening of a subsequent day.

"He kept his appointment, and admitted that the time was favorable. While waiting for my neighbor, before mentioned, to come and witness the performance, we took pipes and coffee, and the magician chatted with me on different subjects. He is a fine, tall, and stout man, of a rather fair complexion, with a dark-brown beard; is shabbily dressed, and generally wears a large green turban, being a descendant of the Prophet. In his conversation he is affable and unaffected. He professed to me that his wonders were effected by the agency of *good* spirits; but to others he has said the reverse—that his magic is satanic.

"In preparing for the experiment of the magic mirror of ink, which, like some other performances of a similar nature, is here termed 'darb elmendel,' the magician first asked me for a reed-pen and ink, a piece of paper, and a pair of scissors, and, having cut off a narrow strip of paper, he wrote upon it certain forms of invocation, together with another charm, by which he professes to accomplish the object of the experiment. He did not attempt to conceal these; and on my asking to give me copies of them he readily consented, and immediately wrote them for me, explaining to me, at the same time, that the object he had in view was accomplished through the influence of the two first words, 'Tarshun' and 'Taryooshun,' which, he said, were the names of two genii, his familiar spirits.

"Having written these, the magician cut off the paper containing the forms of invocation from that upon which the other charms were written, and cut the former into six strips. He then explained to me that the object of the latter charm (which contains part of the twenty-first verse of the Soorat Kaf, or fiftieth chapter of the Kur-an) was to open the boy's eyes in a supernatural manner, to make his sight pierce into what is to us the invisible world. I had prepared, by the magician's direction, some frankincense and coriander-seed, and a chafing-dish with some live charcoal in it. These were now brought into the room, together with a boy, who was to be employed—he had been called in, by my desire, from among some boys in the street returning from a manufactory, and was about eight or nine years of age. In reply to my inquiry respecting the descriptions of persons who could see in the magic mirror of ink, the magician said that they were a boy not arrived at puberty, a virgin, a black female slave, and a pregnant woman. The chafing-dish was placed before him and the boy, and the latter was placed on a seat. The magician now desired my servant to put some frankincense and coriander-seed into the chafing-dish; then, taking hold of the boy's right hand, he drew in the palm of it a magic square, of which a copy is here given:

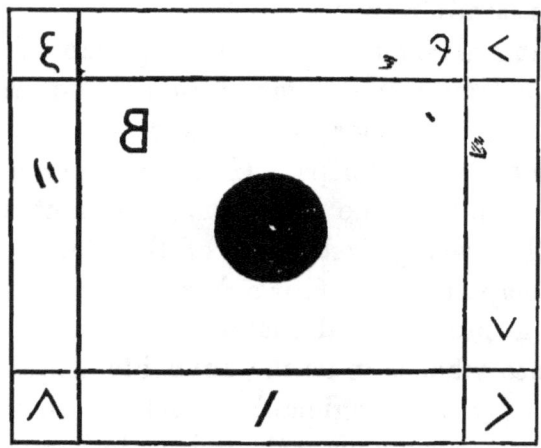

"The figures which it contains are Arabic numerals. In the centre he poured a little ink, and desired to look into it and tell him if he could see his face reflected in it. He replied that he saw his face clearly. The magician, holding the boy's hand all the while, told him to continue looking intently into the ink, and not to raise his head.

"He then took one of the little strips of paper inscribed with the forms of invocation, and dropped it into the chafing-dish upon the burning coals and perfumes, which had already filled the room with their smoke, and as he did this, he commenced an indistinct muttering of words, which he continued during the whole process, excepting when he had to ask the boy a question, or to tell him what he was to say. The piece of paper containing the words from the Kur-an he placed inside of the boy's takee-yeh, or skull-cap. He then asked him if he saw anything in the ink? and was answered 'No;' but about a minute after, the boy, trembling and seemingly much frightened, said: 'I see a man sweeping the ground!' 'When he has done sweeping,' said the magician, 'tell me.' Presently, the boy said: 'He has done.' The magician then again interrupted his mutterings to ask the boy if he knew what a 'beyrak' (or flag) was; and being answered 'Yes,' desired him to say, 'Bring a flag.' The boy did so; and soon said: 'He has brought a flag.' 'What color is it?' asked the magician. The boy replied, 'Red.' He was told to call for another flag; which he did; and soon after he said that he saw another brought, and that it was black. In like manner, he was told to call for a third, fourth, fifth, sixth, and seventh, which he described as being successively brought before him, specifying their colors, —white, green, black, red, and blue. The magician then asked him, (as he did also each time that a new flag was described as being brought,) 'How many flags have you now before you?' 'Seven,' answered the boy. While this was going on, the magician put the second and third of the small strips of paper upon which the forms of invocation were written into the chafing-dish; and fresh frankincense and coriander-seed having been repeatedly added, the fumes became painful to the eyes. When the boy had described the seven flags as appearing to him, he was desired to say, 'Bring the Sultan's tent, and pitch it.' This he did; and in about a minute after he said: 'Some men have brought the tent—a large, green tent—they are pitching it;' and presently added: 'They have set it up.' 'Now,' said the magician, 'order the soldiers to come, and to pitch their camp around the tent of the Sultan.' The boy did as he was desired, and immediately said: 'I see a great many soldiers, with their tents; they have pitched their tents.' He was then told to order that the soldiers should be drawn up in ranks; and having done so, he presently said that he saw them thus arranged. The magician had put the fourth of the little strips of paper into the chafing-dish, and soon after did the same with the fifth. He now said: 'Tell some of the people to bring a bull.' The boy gave the order required, and said: 'I see a bull; it is red; four men are dragging it along, and three are beating it.' He was told to desire them to kill it, and cut it up, and to put the meat into saucepans and cook it. He did as he was directed, and described these operations as apparently performed before his eyes. 'Tell the soldiers,' said the magician, 'to eat it.' The boy did so, and said: 'They are eating it. They have done, and are washing their hands.' The magician then told him to call for the Sultan; and the boy having done this, said; 'I see the Sultan riding to his tent on a bay horse, and he has on his head a high red cap; he has alighted at his tent, and sat down within it.' 'Desire them to bring coffee to the Sultan,' said the magician, 'and to form the court.' These orders were given by the boy, and he said he saw them performed. The magician had put the last of the six little strips of paper into the chafing dish. In his mutterings I distinguished nothing but the words of the written invocation, frequently repeated, excepting on two or three occasions, when I heard him say, 'If they demand information, inform them; and be ye veracious.' But much that he repeated was inaudible; and as I did not ask him to teach me his art, I do not pretend to assert that I am fully acquainted with his invocations.

"He now addressed himself to me, and asked me if I wished the boy to see any person who was absent, or dead. I named Lord Nelson, of whom the boy had evidently never heard; for it was with much difficulty that he pronounced the name, after several trials. The magician desired the boy to say to the Sultan, 'My master salutes thee, and desires thee to bring Lord Nelson—bring him before my eyes, that I may see him, speedily.' The boy said so, and almost immediately added: 'A messenger has gone and has returned, and brought a man dressed in a black suit of European clothes; the man has lost his left arm.' He then paused for a moment or two, and looking more intently and more closely into the ink, said: 'No; he has not lost his left arm; but it is placed to his breast.' This correction made his description more striking than it had been without it, since Lord Nelson generally had his empty sleeve attached to the breast of his coat; but it was the *right* arm that he had lost. Without saying that I suspected the boy had made a mistake, I asked the magician whether the object appeared in the ink as if actually before the eyes, or as in a glass, which makes the right appear the left. He answered, 'That they appeared as in a mirror.' This rendered the boy's description faultless.

"The next person I called for was a native of Egypt, who has been for many years a resident in England, where he has adopted our dress, and who had long been confined to his bed by illness before I embarked for this country. I thought that his name, one not very uncommon in Egypt, might make the boy describe him incorrectly; though another boy, on a former visit of the magician, had described this same person as wearing a European dress, like that in which I last saw him. In the present case the boy said, 'Here is a man brought on a kind of bier, and wrapped up in a sheet.' This description would suit supposing the person in question to be still confined to his bed, or if he be dead. The boy described his face as covered, and was told to order that it should be uncovered. This he did, and said, "His face is pale, and he has mustaches, but no beard,' which was correct.

"Several other persons were successively called for,

but the boy's descriptions of them were imperfect, though not incorrect. He represented each object as appearing less distinct than the preceding one, as if his sight was gradually becoming dim; he was a minute or more before he could give any account of the persons he professed to see towards the close of the performance; and the magician said it was useless to proceed with him.

"Another boy was then brought in, and the magic square, etc., made in his hand; but he could see nothing. The magician said he was too old.

"Though completely puzzled, I was somewhat disappointed with his performances, for they fell short of what he had accomplished in many instances in presence of certain of my friends and countrymen. On one of these occasions an Englishman present ridiculed the performance, and said that nothing would satisfy him but a correct description of the appearance of his own father, of whom he was sure no one of the company had any knowledge. The boy accordingly having called by name for the person alluded to, described a man in a Frank dress, with his hand placed to his head, wearing spectacles, and with one foot on the ground and the other raised behind him, as if he were stepping down from a seat. The description was exactly true in every respect; the peculiar position of the hand was occasioned by an almost constant headache, and that of the foot or leg by a stiff knee, caused by a fall from a horse in hunting. I am assured that on this occasion the boy accurately described each person and thing that was called for.

"On another occasion Shakspeare was described with the most minute correctness, both as to person and dress; and I might add several other cases in which the same magician has excited astonishment in the sober minds of Englishmen of my acquaintance. A short time since, after performing in the usual manner by means of a boy, he prepared the magic mirror in the hand of a young lady, who, on looking into it for a little while, said that she saw a broom sweeping the ground without any body holding it, and was so much frightened that she would look no longer."

From the above minute account, it is very evident that the different boys absolutely did see the images of the persons and things which, in most instances, they so correctly described, but it is also very apparent that they sometimes also imagined, and saw what they imagined, as is sometimes the case with subjects who look at things in a state of Artificial Somnambulism.

From the same account it is also very plain that the magician himself was ignorant of its true nature; and, although artful and calculating in his movements, was no doubt only following the routine or prescribed method of those from whom he had learned the mystery.

I cannot subscribe to the idea that the frankincense or the coriander-seed had anything to do in producing the effects, any more than the slips of paper with the written names of the genii upon them, the square, or the Arabic numerals therein placed, etc.

It is very certain, too, that looking intently at the ink alone, without the magic square or any of the other mummeries, would have had the same effect upon the boys' eyes, and their "trembling" and seeming "frightened," proves that they were entering another state; and as they generally saw nothing before this took place, although asked, it renders the probability more conclusive.

All the desires, questions, and commands, etc., about the flags, the Sultan, the soldiers, the pitching of the tents, and the killing and the eating of the bull, etc., were only so many artful or necessary ways of gaining time to ascertain whether the boys could see correctly—these things having all been imagined or thought of by the magician, and gotten from his mind by the boys, as they did also the image of Lord Nelson, Shakspeare, and the rest from those of the persons who named them respectively and desired a description of them.

I have had subjects in a state of Artificial Somnambulism to do these things hundreds of times, and it is nothing more than clairvoyance or reading the mind, which, in the cases of the boys above described, I conceive was brought about in the following manner:

In the first place, the magician addresses the individual and desires to know what he wishes to have brought before the eyes of the boy. His doing this naturally draws the boy's attention to the mind of the individual who is about to make the request, and as the individual utters the name he naturally forms the image of the person he desires to be called and seen by the boy in his own mind, and keeps the same image there more or less during the experiment. This image so formed, the boy, being clairvoyant, would most naturally see and describe correctly, according as *that image* was formed *perfectly* in the mind of the individual, or as he (the boy) had looked without imagining.

But it may be said that the boy did not enter the state of Artificial Somnambulism entirely, and therefore could not have been clairvoyant.

To this objection I reply that I have long since proved that it is possible for any part of the body to enter this state independent of the rest; consequently the eye of the boy, or rather the necessary portion of the brain, may still have been in this condition. It is true the boys had their eyes open, but I must here also anticipate this objection by stating that it is possible for any one to enter this state with the eyes open as well as with them shut; and I have many to do so at pleasure, as the following conclusion of a letter from myself to the editor of the "Magnet," and published in that journal, will illustrate. In that letter I stated that experiments proved that clairvoyance or *mind sight* was entirely different and superior to the sight of the natural eye, and that the many facts which were presented from day to day induced me to believe that the sense of sight or the natural eye could also be used at the will of the subject while in a somnambulic condition, if the lids were to remain open, independent of the mind's sight—clairvoyance—which they heretofore only used when the lids were closed.

Accordingly I requested Miss ——— to keep her eyes open while she entered the sleep. This she readily accomplished, and in about one minute she was in a perfect state of Artificial Somnambulism with the eyes open. The facts elicited were as follows: First, that she was enabled to see with either the mind or the natural eye as she felt disposed, but could not use both at one and the same time. Secondly, when she looked with the eye it had a natural appearance, but when she looked with her mind, or clairvoyantly, the eye assumed a heavy or drowsy appearance. Thirdly, she could see me or anything else through the wall when she looked clairvoyantly, but could not when she looked with the natural eyes.

Since that time I have had many others in this state with the eyes open, and have been enabled to perform many interesting experiments. Their appearance is so perfectly natural when they look with the natural eyes, that I have had several, with proper instructions, to enter a room and to converse freely with a number of ladies and gentlemen, without any one noticing that they were in this state until the fact was mentioned and the case explained.

The above facts not only prove that they can have

their eyes open while in this state, but that they can also be in it perfectly without its being noticed by those who do not examine into their condition particularly, and there can be no question but that the same is the case with those who are practised upon by the Egyptian magicians.

It is also very evident that the power of vision in the boys under the care of the Egyptian magicians is identical with and is nothing more than the same power of reading the mind, which subjects in a state of Artificial Somnambulism possess, and are simply the ideas or pictures formed in the minds of those who have made such requests. And when the boys do not imagine anything themselves, they have always described the persons exactly as the individual who made the request had pictured those persons in his mind, with their clothes on, and the peculiarities which belonged to them when they were alive.

III.—OF THE "MYSTERIOUS LADY."

The "Mysterious Lady," who exhibited her clairvoyant powers, etc., in the various cities, was also in a state of Artificial Somnambulism, with her eyes open; and, although her experiments were very successful, they were, by no means, more so than any other good clairvoyant.

IV.—OF THE EARTH MIRRORS.

The earth mirror, so called by some, because it was supposed that certain persons, born in the month of December of any year, by taking it into a dark room and fixing their eyes intently upon it for some time, could see treasures, etc., which may have been buried in the earth.

I am acquainted with a number of persons who believe most reverentially in the virtue of this glass, and whom all the philosophical arguments in the world would not convince to the contrary; because they have seen persons who have (as they say), by looking into it, observed certain things at a distance which really took place, or were found to correspond to their statements, although they were entirely ignorant of the facts before they were desired to look for them. The virtue is entirely ascribed to the glass, and it is stated that some who have looked into it, frequently become so that they can look into it in "broad daylight," and see as well as they can when they take it into a dark room.

This is nothing more than another method of throwing the "mind's eye" into the somnambulic state, rendering it clairvoyant, and I am persuaded that the same persons could do the same thing as well by looking at anything else, particularly if something else were substituted without their knowledge.

I have seen several kinds of these glasses, and give on the next page an outline of two which I obtained. The first consists of a piece of common looking-glass about four inches square, fastened upon a piece of wood about the same size. Upon its surface is scratched two circles, the one within the other. In the inside of these is a double triangle, also one within the other, with several strange names at various distances, both outside and inside of the circles and triangles.

FIRST EARTH GLASS.

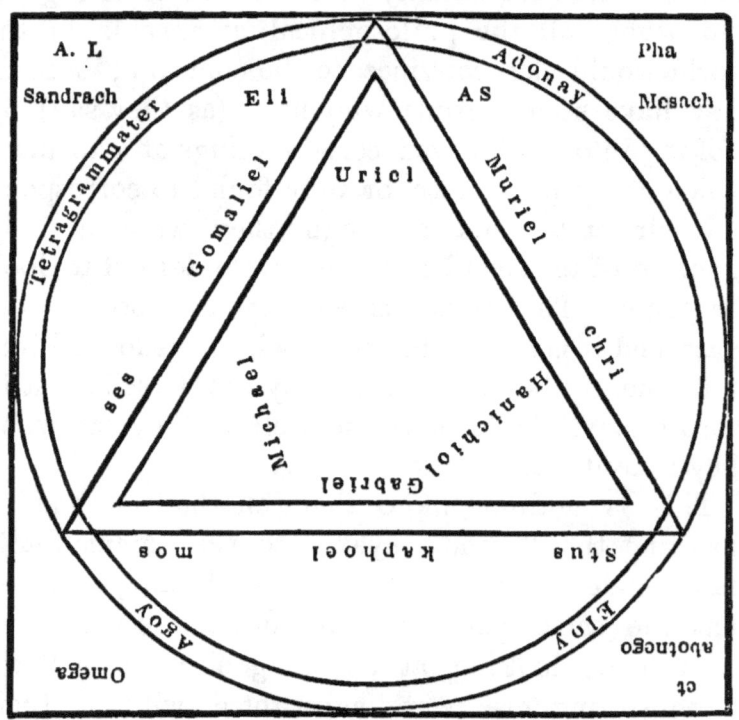

The second is much smaller, and consists of a plain piece of glass, on the surface of which is also scratched two circles, the one within the other, and two double triangles, the planes of two crossing the planes of the other two, with various letters, characters, and numbers placed around and between both.

SECOND EARTH GLASS.

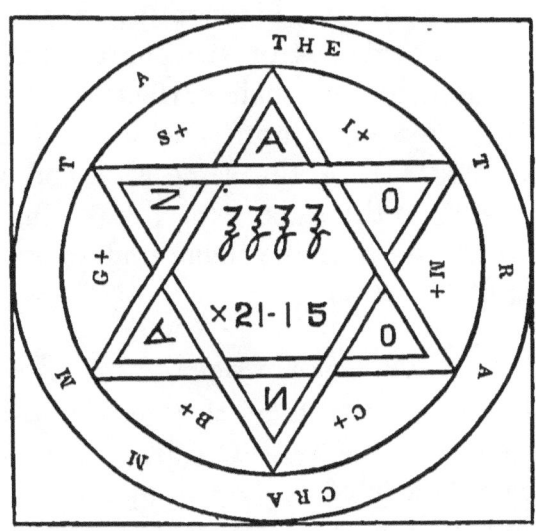

The second earth glass is considered of singular virtue, not only on account of the nature of its characters, etc., but because it is enclosed in a box made of wood which was once the bottom of a coffin. In this box, and beneath the glass, is placed a slip of paper upon which is written the first chapter of Genesis, together with the symbols of the various planets and other characters.

Here mystery and superstition are again hand in hand; but it is very certain that, looking steadily and intently at anything else in the dark, would answer the same purpose, because we cannot see either; and it is only the fixing of the eye, together with the intense looking and *the desire to see*, which produces the effects which follow.

It is also very clear that the month in which the person who looks, is born, has nothing to do with it any more than the glasses, or the characters, etc., engraven or scratched upon their surfaces.

Many other mysteries, both ancient and modern, can be accounted for in the same way, and are undoubtedly owing to the same cause, requiring only a proper investigation to make them understood. The following is of this stamp:

V.—SECOND SIGHT.

Many of the inhabitants of the western islands of Scotland are said to be possessed of this power. It is said to be a faculty of seeing things which are to come, or things done at a distance, and represented to them as if actually visible and present.

An account of it is given in Johnson's journey to these islands, and it is said, that "if a man be dying, or about to die, his image appears to them in its natural shape, etc., although they never saw his face before; and it is believed, by some, that when a person be so seen by them, if he be not dead he will certainly die.

"The power is also said to be hereditary, nor can the person exercise it at pleasure, or communicate it to another, but comes on him involuntarily and arbitrarily, often to his great trouble and terror, and is held so much in discredit that none will counterfeit it, but, on the contrary, dissemble and conceal it."

This is also nothing more than partial somnambulism, and the "mind's eye" has fallen into this state naturally, or through the belief or fear on the part of the persons themselves. Their powers, however, are also only those of other clairvoyants, and if they knew the true nature of the state, could so exercise them. I am acquainted with a number of persons who can exercise this power at pleasure; but in no case *can they, or any others, see or unerringly foretell* that which is yet to come. They may guess, and sometimes be correct, as when awake. It is true they can see and tell what is passing at the time of looking, at any distance; but not what is yet to transpire. Certain portions of the brain may at times become perceptive, or clairvoyant, and, as it were, reach out to a distance, independent of our knowledge; and we may, in such cases, have knowledge or presentiment of certain things happening at a distance, long before any *positive* information can be received through the usual channels of communication. Second-sight is the same kind of clairvoyance, or foresight, and nothing more. The future, so far as my experience goes, can only be arrived at by impression, induction, or reasoning from events that have taken place, and not by positive seeing or knowledge.

VI.—PHANTASMS.

Phantasms are entirely different from second-sight. Phantasms are caused by fevers, or mental excitements of any kind; and, in the case of M. Nicoli, described by himself, and quoted in Mr. Sunderland's work on Pathetism as a case of second-sight, is one purely of this character, and brought about by "an almost uninterrupted series of misfortunes," producing a derangement of the cerebral functions of one side of the head.

I quote below the same case from the above work as one of phantasm purely, and entirely different from second-sight. The account was drawn up by M. Nicoli himself, and is as follows:

"During the latter ten months of the year 1790, I had experienced several melancholy events, which affected me, particularly in September, from which time I suffered an almost uninterrupted series of misfortunes, which afflicted me with the most poignant grief. I was accustomed to be bled twice a year; this was done on the 9th of July, but was omitted to be repeated at the end of the year. Less blood had consequently been evacuated in 1790 than was usual with me; and from September I was constantly occupied by business, which required the most unremitting exertion, and which was rendered still more perplexing by frequent interruptions.

"In January and February of the year 1791, I had the additional misfortune to experience several unpleasant circumstances, which were followed on the 24th of February by a most violent altercation. My wife and another person came into my apartment at ten o'clock in the morning, in order to console me, but I was too much agitated by a series of incidents which had most powerfully affected my moral feelings, to be capable of attending to them. On a sudden, I perceived, at a distance of about ten paces, a form like that of a deceased person. I pointed at it, asking my wife whether she did not see it? It was but natural that she should not see anything; my question, therefore, alarmed her much, and she sent directly for a physician. The phantom continued about eight minutes. I grew at length more calm, and being extremely exhausted, fell into a restless slumber, which lasted about half an hour. The physician ascribed the apparition to violent mental excitement, and hoped there would be no return; but the violent agitation of my mind had in some way disordered my nerves, and produced further consequences, which deserve a more minute description. At four o'clock in the afternoon, the form which I had seen in the morning reappeared. I was by myself when this happened, and, being rather uneasy at the incident, went to my wife's apartment, but there, likewise, I was followed by the apparition, which, however, disappeared at intervals, and always presented itself in a standing posture. About six o'clock, there appeared also walking figures, which had no connection with the first.

"I cannot assign any other cause for all this, than a continued rumination on the vexations I had endured, which, though calmer, I could not forget, and the consequences of which I meditated to counteract. These agitations occupied my mind three hours after dinner, just when digestion commenced. I consoled myself at length with respect to the disagreeable incident which had occasioned the first apparition; but the phantasms continued to increase and change in the most singular manner, though I had taken the proper medicines, and found myself perfectly well.

"When the first terror was over, as I beheld these phantasms without great emotion, whilst taking them for what they really were—the remarkable consequences of an indisposition, I endeavored to collect myself as much as possible, that I might preserve a clear consciousness of the changes that should inwardly take place in me. I observed these phantasms very closely, and frequently reflected on my antecedent thoughts to discover, if possible, by means of what association exactly these forms presented themselves to my imagination. I thought at times I had found a clue; but, taking the whole together, I could not make out any natural connection between the state of my mind, my occupations, train of thoughts, and the multifarious forms which now appeared to me, then again disappeared. After repeated and close observations, and a calm examination, I was unable to form any conclusion relative to the origin and duration of the different phantasms which presented themselves to me. All that I could infer was, that while my nervous system was in such an irregular state, such phantasms would appear to me as if I actually saw and heard them; that these illusions were not even modified by any known laws of reason, imagination, or the common association of ideas; and that probably other people, who may have had similar apparitions, were exactly in the same predicament. The origin of the individual forms which appeared to me were undoubtedly founded on the state of my mind; but the manner in which it was thus affected will probably remain as inscrutable as the origin of thought and reflection.

"After the first day, the form of the deceased person no longer appeared, but in its place many other phantasms, sometimes representing acquaintances, but mostly strangers. Those whom I knew consisted of both living and dead persons; but the number of the latter was comparatively small. I observed that persons with whom I daily conversed did not appear as phantasms; these representing chiefly persons who lived at some distance from me. I attempted to produce at pleasure phantasms of persons whom I knew, by attentively reflecting on their countenance, shape, etc., but, distinctly as I recalled to my lively imagination the shapes of those persons, still I labored in vain to make them appear to me as phantasms, though I had before involuntarily seen them in that manner, and perceived them some time after, when I least thought of them. These phantasms appeared to me contrary to my inclination, as if they were presented to me from without, like the phenomena of nature, though they existed nowhere but in the mind. I could, at the same time, plainly distinguish between phantasms and real objects; and the calmness with which I examined them enabled me to avoid committing the smallest mistake. I knew it exactly when it only appeared to me, that the door was opening and a phantasm entering the room, and when it actually opened and a real person entered. The phantasms appeared to me equally clear and distinct at all times, and under all circumstances, both when I was alone and when I was in company, as well in the day as at night, and in my own house as well as abroad. They were, however, less frequent when I was in the house of a friend, and rarely appeared to me when in the street. When I shut my eyes, these phantasms would sometimes disappear entirely, though there were instances when I beheld them with my eyes closed; yet, when they disappeared on such occasions, they gradually reappeared when I again opened my eyes. I conversed occasionally with the physician and my wife respecting the phantasms which surrounded me at the moment. They appeared more frequently walking than at rest; nor were they constantly present. They frequently did not appear for some time; but always reappeared for a longer or shorter period, either singly or in company; the latter, however, was most often the case.

"I generally saw human forms of both sexes, but they usually seemed not to take the smallest notice of each other, moving as in a market place, where all are eager to press through the crowd. At times, however, they seemed to be transacting business with each other. I also repeatedly saw people on horseback, dogs, and birds. All these phantasms appeared to me in their natural size, and as distinct as if alive, exhibiting different shades of carnation in the uncovered parts, as well as different colors and fashions of their dress, though the colors seemed to me to be paler than in real nature. None of the figures appeared particularly terrible, comical, or disgusting; most of them being of an indifferent shape, and some having a pleasing appearance. The longer these phantasms continued to appear the more frequently did they return, whilst at the same time they increased in number.

"About four weeks after their first appearance, I began also to hear them speak. They sometimes conversed among themselves, but more frequently they directed their discourse to me. Their speeches were commonly short, and never of an unpleasant tenor. Several times I saw beloved and sensible friends of both sexes, whose addresses tended to appease my grief, which had not wholly subsided. These consolatory speeches were in general addressed to me when I was alone; sometimes, however, I was accosted by these consoling friends whilst in company, even while real persons were speaking to me. These consolatory addresses consisted sometimes of abrupt phrases, and at other times they were regularly connected.

"Though both my mind and body were in a tolerable state of sanity at this time, these phantasms became so familiar to me that they did not cause me the slightest uneasiness.

"I even sometimes amused myself with surveying them, and spoke jocularly of them to the physician and my wife, yet I did not neglect to use the proper medicines, especially when they began to haunt me the whole day and even at night as soon as I awoke. At last it was agreed that leeches should again be applied to me, as formerly, which was accordingly done on the 20th of April, 1791, at eleven o'clock in the morning. No one was with me besides the surgeon, but during the operation my chamber was crowded with human phantasms of all descriptions. This continued without interruption till about half-past four, when my digestion commenced. I then perceived that they began to move more slowly; soon after their colors began to fade, and at seven o'clock they were entirely white, and moved very little, though the forms were as distinct as before, growing however by degrees more obscure, yet not fewer in number as had generally been the case. The phantasms did not withdraw nor did they vanish, which previous to that time had frequently occurred. They now seemed to dissolve in the air, while fragments of them continued visible a considerable time. About eight o'clock the room was entirely cleared of my fantastic visitors. Since that period I have felt twice or three times a sensation as if these phantasms were going to reappear, without, however, actually seeing anything. The same sensation surprised me just before I drew up this account, whilst I was examining some papers relative to these apparitions which I had drawn up in the year 1791."

It is very evident from the above minute account that what was seen by M. Nicoli, were phantasms simply, and were produced by a determination of blood to certain portions of his brain. Persons laboring under fevers are often similarly affected, and in aggravated cases produce delirium, which is only a more general and a greater degree of the same affection. Phantasms, as I have said, are caused by an undue excitement, from whatever cause, of the functions of the imagination and perception in the faculties of Seeing, Hearing, Individuality, Form, Size, Color, Configuration, Language, etc., independent of the functions of the will in the same faculties, and is entirely different from second-sight, which recognizes things and persons that really exist, and circumstances that are transpiring at the time in the outer world. M. Nicoli saw involuntarily, and could not observe anything he might desire, although the same persons and things were seen by him involuntarily.

Those who possess the power of second-sight I am persuaded could, if they knew the nature of their condition, use their clairvoyant powers as well as those who are in a somnambulic state, and instead of its coming to them involuntarily, they could use the power at pleasure.

CHAPTER XII.

TRANSPOSITION OF THE SENSES.

THE transposition of the senses to the pit of the stomach and other parts of the body in cases of what has been termed Catalepsy, has been witnessed and recorded by men of the first standing and abilities, and is now almost universally believed to be true.

The descriptions given of it do not differ materially, and when a person is "suddenly seized," as is usually the case, "the senses and power of voluntary motion are as suddenly suspended; so that the patient remains fixed in the position in which he happens to be at the moment of seizure, and the limbs commonly retain any position into which they are put by external force. During the paroxysm the sensations are in general suspended; the patient neither receives any impressions from external objects, nor retains any recollection of what happened during the fit.

"The vital functions continue to be performed, but more feebly; the pulse and respirations are regular, but the former is smaller than in health; the color of the countenance usually undergoes little or no change. After a duration, which varies in different instances, commonly after a few minutes, sometimes after a lapse of a few hours, and occasionally, though rarely, after a continuance of three or four days, the paroxysm suddenly declines. The patient awakes, as it were, from a sleep, generally with deep sighing, and all the functions of the body are restored."

These symptoms constitute what has been called by systematic writers a "perfect paroxysm." Yet many of these symptoms vary in different persons, and the following account of Mademoiselle Malanie by Dr. Durand, of Caen, has been quoted by many medical writers and published in many other works, both in Europe and this country. I here extract a part of the original account, which is as follows:

"Six days after the first attack of hysteria the patient became suddenly dumb, and continued so for three days, being unable to articulate a single word: on the fourth day she recovered the power of speech at the termination of a severe hysterical attack; the surprise, however, expressed by those about her at hearing her speak, threw her into a fresh fit, which lasted for three hours and ended in catalepsy; this was on the 30th of August, 1841. From this period the patient was seized every day with several attacks of catalepsy alternating with hysteria, and lasting about half an hour.

"During the cataleptic accesses there was complete insensibility of every part of the body; the limbs remained in the most fatiguing positions without stirring, the respiratory movements were imperceptible, and the pulsations of the heart, which could scarcely be felt, were from sixty to seventy per minute. After a few days the cataleptic fits became longer, and lasted for several hours, being, however, occasionally interrupted for a minute or two whenever the girl coughed. Sometimes she would turn round in her bed, or sit up; at others she would suddenly start up, without opening her eyes, and place herself on the edge of the bed, or on some piece of furniture, in a most fatiguing posture; in this state she would remain until a fit of coughing came on, or until she was brought back to her bed. Although the eyes were constantly shut, *she avoided every obstacle* carefully, and seemed heedless of risks which would have alarmed any one in a normal state. On one occasion she left her bed during a fit of coughing, ran to the window and opened it, and before any one could come to her assistance she had one foot out of the window, but the cough suddenly ceased, she became cataleptic, and remained in the same position until some people came and placed her in bed.

" When the fits of hysteria and catalepsy ceased the patient recovered all her faculties, and merely complained of fatigue and her ordinary pain in the side. Five weeks after the first attack of catalepsy Mdle. Malanie fell several times into a state of natural somnambulism. She would get up without opening her eyes, walk about her room, arrange her furniture, and enter into conversation with those about her, often mentioning circumstances which she would have wished to conceal; after remaining in this state for several hours she fell into a state of catalepsy, indicated by apparent suspension of the respiration and complete silence. On the 12th of October, a few days after her first access of somnambulism, I found the patient in a state of catalepsy. Having placed my hand on the epigastric region, I noticed that her countenance became expressive of pain. I then placed my lips on the pit of her stomach, and asked her several questions. To my astonishment she answered correctly; for, although I had read most of the histories of this kind recorded in different works, I did not believe one of them. During the first examination I made numerous experiments, which led me to conclude that there was *a transposition of the five senses to the pit of the stomach.* On the evening of this day I made fresh experiments, during three hours, in the presence of numerous witnesses, who were not less surprised than myself. In a word, during two months, I renewed the experiments daily, and often several times a day, making use of every precaution to avoid deception, and having numerous witnesses around me."

I ascribe the above results to the manner in which the doctor proceeded, and they would have been much more perfect if he had understood her powers. He succeeded in drawing her attention to her stomach, and found that she answered questions correctly, which he, not knowing better, had addressed only to that part. She could have answered the same questions just as well if he had applied his lips to any other part of her body, because all persons in this state possess such powers; and I have seen the same transposition effected at the will of the subject, in many cases of Artificial Somnambulism, by simply requesting them to throw their minds to the stomach or any other part of the body. She could have done the same if she had attempted it, either at his request or of her own accord. But further, he goes on to say—"I shall now relate the results of these experiments. During the cataleptic state the muscles presented three different conditions. Sometimes they were all relaxed, and the limbs could be placed in any position, which they retained, however fatiguing the posture might be; at other times all the muscles were in a state of rigid contraction; at others again, they were relaxed, and the limbs fell down when raised from the body."

These conditions of the muscles, etc., I have also witnessed a thousand times in cases of Artificial Somnambulism, and are natural to this state when the mind of the patient is not placed upon them, and can be assumed by them at pleasure, viz. :—If the arm be raised and they have not their mind expressly upon it, it will remain in that position; simply because it is as easy for them to have it in that position as in any other, and if their attention be called to it, and the question is asked: Why do you hold your arm in that position? they will tell you, with the utmost candor, that they did not know that it was raised; and if they do not, by an express act of their will, change the position it will remain there—no matter how awkward it may seem—for an indefinite period. The doctor also stated, that "there was no sensibility in any part of the body except over the pit of the stomach, the palms of the hands, and the soles of the feet. Thus, we might pinch the skin, or pierce it with pins, or pull out the hair, tickle the nose, etc., without eliciting any sign of feeling. On the contrary, if the pit of the stomach, soles of the feet, or palms of the hands were touched, even with the point of a feather, the girl immediately withdrew the part touched, and her countenance indicated displeasure. When a Leyden jar was placed in communication with the parts just named, she had a violent commotion, or was suddenly awakened, but the jar might be discharged on any other part of the body without producing the slightest effect.

"The ears appeared to be insensible to sound, the loudest noise did not attract her attention, but when a small bell was agitated over the sensitive parts her countenance showed she heard the noise. If the lips were placed in contact with the sensitive parts she heard everything that was said, although the voice was so low that it could not possibly reach the ears.

"Her answers were delivered in an exceedingly low tone, and, generally speaking, the person appointed to catch them would repeat them without hearing the question asked.

"The patient never spoke, except when her limbs

were in a state of relaxation; during the rapid cataleptic state the tongue and organs of speech were immovable.

"The senses of taste and smell were not exercised by their natural organs, but were very acute in the sensitive parts. Thus, we filled the nose with assafœtida or tobacco, placed bottles of ether, concentrated ammonia, etc., under the nose, without producing the least effect; but when a small portion of a sapid body was placed in contact with the sensitive *parts the patient distinguished it at once.* Thus, she recognized and named, one after another, the syrup of poppies, vinegar, gum, and capillaire, wine, water, orange flower water, seidlitz water, currant jelly, etc., although only one or two drops of each substance were placed on the palm of her hand. When a few grains of snuff were placed on the sole of her foot she sneezed at once, and thus easily distinguished French snuff from English snuff.

"Although the result of my first experiments induced me to think the sense of vision was transposed as well as the other senses, subsequent trials showed that what I had regarded as vision was nothing more than an exquisite sense of touch. When an object was placed on any of the *sensitive parts*, and she was asked if she saw it, she answered 'yes,' and immediately named the object if she was acquainted with it; or if not, gave a correct description of the body. Thus, she always detected a watch if placed over the pit of the stomach, and never failed to tell whether it was made of gold or silver, going or stopping. If asked the hour she would answer pretty correctly as to the *true* time of day; but if the hands of the watch were designedly changed she always failed to tell the time they marked. She could distinguish and name every kind of French coin placed in her hand, but not the name of the sovereign in whose reign they were struck; she could distinguish a bit of silk from a bit of cloth, but not their respective colors.

"At the second sitting she succeeded in spelling the word *commerce*, written in large letters, and placed upon the pit of her stomach; this required considerable effort, and she complained for a long time of fatigue. In subsequent experiments, however, she was never able to distinguish any of the letters of the alphabet, when placed in contact with the sensitive parts."

I have seen all the above peculiarities in persons who were in a state of Artificial Somnambulism, and have performed similar experiments with many of them, not only in the above manner, but also by placing things on various other parts of their bodies and directing their minds to them. The experiments succeeded just as well when the articles were placed at a distance as when in actual contact with their bodies.

If the snuff which was placed to the soles of Mademoiselle Malanie's feet had been taken to the next room and her mind *properly* directed to it, she could have distinguished between the French and the English kinds as well as if placed on any of what he supposed to be the *only sensitive parts.*

Her failing, in many of the experiments made to test her powers of vision, was owing to the doctor not drawing her attention to them properly, or a carelessness or unwillingness on her part to do them at all; and I am persuaded that, if she had been properly directed, or had been perfectly willing on her own accord, she could have done so correctly as well at a distance as if placed in actual contact with her body.

I can therefore see no essential difference in what is called catalepsy and that which I denominate Artificial Somnambulism. The first is fallen into accidentally, and the subjects are ignorant of their powers, or their condition, in which accidental or hysterical causes have placed them; while the latter is induced by instructions or the will of the subject, with a full knowledge of the condition and their powers therein.

Her becoming rigid at one time, and relaxed at another, was an act of her own will, or because she thought it must be so; but if she had known her powers, she could have caused or prevented it at pleasure. She could also have translated her sensibility, etc., to any other part of her body as well as to her stomach, the palms of her hands, and the soles of her feet, and could have rendered the most sensible parts insensible, and performed all the above experiments at any other point at pleasure.

What has been related of her hearing at these points would at first appear ridiculous, and it would be very natural to suppose that if she heard a small bell, or even a whisper, at any part of her body, that she could also hear a very loud noise made in the same room. The reason, however, why she did not, was because she *believed* that she could not, and paid no attention to anything that was not held to those parts; and is upon the principle that they hear, see, feel, taste, or smell nothing but that which their attention is placed upon; as is the case also with most persons in a state of Artificial Somnambulism.

I have seen but one genuine case of what has been termed catalepsy. This was in a girl about seventeen years of age, and was brought on by repeated attacks of hysteria, produced, in the first instance, by a dread of witches, which her grandmother made her believe were afflicting her.

While in this condition, she was perfectly rigid and motionless, with an *apparent* suspension of all the senses. I relieved her, however, instantly, by stating *aloud* (so that she might hear it, if paying attention to me) that I could do so by placing my hand upon her forehead. I no sooner had done so than she awoke, and, although somewhat confused in her ideas, and having a silly expression, she gradually recovered, and has since remained entirely free from it. This case was identical with those of catalepsy, and very much resembled the condition which is often witnessed and produced by religious excitement.

I relieved her because she *believed* that I could. If she had not had that belief, or had not been paying attention to me, I might have worked with her for hours, or even days, without effect, as those who attended her had frequently done before. Such is the power of mind in this state, not only over affections of this kind, but many others which are usually considered diseases. The power or influence of the mind over like diseases in our waking moments has been much overlooked, and the influence is an hundred fold greater when in a state of Artificial Somnambulism. I have lately seen many affections, which were considered incurable, yield to the proper direction of the subject's mind while in a somnambulic state, after long and skilful courses of medicine had entirely failed. It is not the imagination, however, which produces these effects, but a fixed and determined resolution, on the part of the subjects, that it shall be otherwise when they awake. But I shall treat of this more fully hereafter.

CHAPTER XIII.

NATURAL SLEEP.

SLEEP is the more or less complete suspension of the faculties of the body; and, when perfect, the person is without sensation, volition, action, or consciousness, and the portion of existence passed in this state is a perfect blank. It is well known, however, that sleep is not always perfect, and, as I have before stated, that it was possible for one or more of the senses, functions, or faculties, etc., to enter the somnambulic state; or, when all were in this condition, it was possible for one or the other of the same functions, etc., to be awakened independent of the rest. This is also the case with the same functions, etc., in a natural sleep; but the awaking or putting to sleep cannot be accomplished by an act of the subject's will, as in the former case, and is usually the result of fatigue or wakefulness in those faculties, etc., which are so affected. When one or more of the functions awake, dreaming ensues, and the mental conceptions, etc., are incongruous, or otherwise, according to the number and nature of the functions and faculties which may be awake, etc.

CHAPTER XIV.

NATURAL SOMNAMBULISM.

MANY authors have given descriptions of this state, in which the subjects are said to have performed various feats unconsciously, or at least knew nothing about them when awoke; such as getting up, opening the doors and windows, going to the barn and harnessing horses, crossing bridges, or even walking on the tops or eaves of houses, and various other feats, for a detailed account of which I must refer the reader to the many accounts already published.

That it is possible for some to fall into a state of this kind is now generally believed to be true; and the circumstances which have happened and the feats which some have performed while in it are not now doubted by those who have made the subject their study. It would seem that it is as natural for some to fall into this state as it is for others to fall into a natural sleep; but it is impossible for any one to tell why they fall into one or the other; and it is a question in my mind whether we may not *all* have been in it at times without our having been conscious of it when we awoke.

It is very certain, however, that before any one

can enter either of the sleeps that they must forget themselves.

This is, as I have before stated, also the case with those who enter the sleeping condition of the artificial state; and those who are enabled to forget themselves quickly, for the most part enter the state readily. But losing themselves very soon is not all that is necessary, for some do this at once, yet fall into a natural sleep.

Those, however, who are self-possessed, and can abstract or withdraw their minds from their surroundings, and can fix it determinately and confidentially upon the result, commonly enter the artificial state with the most facility.

When a person has once been in a state of natural somnambulism, it may or may not become a habit; and I have known some to be cured of the habit by flagellation, fright, or the application of cold water. In these cases, the fear of these measures operate against falling into it, and shows that, when there is anything upon the mind of the subject, that it operates as much against entering it in a natural way as it does in the artificial.

The phenomena in this state are precisely like those in the artificial, with the exception of what is caused by the manner of entering it. That is, in a natural state, they do not generally converse with any one, and commonly awake when spoken to; while in the artificial, if their mind is otherwise engaged, they hear and speak to the person into whose care they have entrusted themselves, and no one else. The reason why they hear him while in an artificial state is, because it is expected or has been agreed upon to do so before they entered it, and their attention while in it is fixed upon him; yet they can hear and speak to any one else whenever they please, which I believe could also be done by those in a natural state if they knew their powers.

I have seen persons in the artificial state to resemble every variety of natural somnambulism from a mere inability to move to catalepsy or the most perfect states, and have no doubt that they are one and the same state, varying only in degree or attended by peculiarities which are constitutional, or have been caused by *a belief* that *they must be so*, whether that belief has been instilled by others or has originated with themselves.

I.—TRANCE.

Trance has long been supposed to be a state in which the soul has left the body and passed into regions of celestial beings, or wrapped in visions of future and distant things, etc.

As I have never seen a case of trance, properly so called, I cannot say that I understand what is the true nature of the condition, but presume it to be a somnambulic state, in which the subject for a certain period remains in an unconscious condition, or roves about in space, remembering what has transpired or not, as the functions of memory, etc., in the various organs, were active or not, and awaking when the mind returned. The following is a case in point with my explanation, which was published in the *Daily Enquirer*, of Cincinnati, May 15th, 1869.

From the South Bend (Ind.) National Union.
AN EXTRAORDINARY CASE—ALLEGED WITCHCRAFT—A YOUNG GIRL IN A TRANCE ELEVEN WEEKS—QUEER REVELATIONS.

Some weeks ago it was our intention to lay the facts of the following very singular case before our people, but at the request of the physician, Dr. Fletcher, have been waiting for further developments.

North of the village of Mishewaka lives a well-to-do farmer named Jacob Martin, a Pennsylvania Dutchman. Mrs. Martin was born in Germany, but the language used in the family for years has been English, the parents preferring to have the children speak that instead of the German language. Mr. M.'s mother had formerly lived in the family and slept with Julia, a girl about thirteen years of age. According to the statements of several who have visited the family, the old lady has for some time had the unenviable reputation of being a witch. About a year ago it was not found agreeable to have her in the family, so she was removed to Mishewaka. Shortly after, Julia was attacked with rheumatism, and in a short time her limbs swelled so that they had to be bandaged. She had to be cared for like an infant from this time until about ten weeks ago, when she fell into a trance, which lasted for three days, during which time she took no nourishment—apparently in a deep sleep, from which no one could succeed in waking her. Several physicians visited her, but could not determine what the symptoms betokened. Dr. Fletcher, of our city, was at this time called, who pronounced it catalepsy, caused by spinal difficulty. On the evening of the third day she awoke in convulsions, during which time she bit her tongue so that a spoon, covered with cloth, had to be kept in her mouth. Several thicknesses of the cloth were bitten through in a few minutes. During a lucid interval she made startling revelations in regard to her grandmother—asserting that she had bewitched her, and was endeavoring to make her chew her tongue out, to prevent her telling the strange things which had been revealed in the visions. Her conversation was carried on in "High" German, which had never been spoken in the family, and much of which they could not understand. When they did not comprehend her, she talked the Pennsylvania Dutch quite as fluently, although she never was able to speak a connected sentence in either language before. She also related a circumstance regarding some ladies of South Bend. Upon investigation it was found to be reported correctly. Told of some persons who were on the way to see her, but had met with an accident, having "broke down in the mud." Shortly after, the ladies arrived and stated that they "were detained by the breaking of their vehicle near South Bend."

On Sunday last she was conscious about an hour,

at which time she insisted that it was the influence of her grandmother which caused her illness. She wept bitterly when she spoke of going away again; said she feared she might never see them again, but stated that she "would be back again on Wednesday, if permitted to return." When asked if she did not want something to eat, she stated that she had eaten, although she had taken no nourishment for several days. When questioned as to the kind of food, she said, "We eat manna where I've been."

On Wednesday, about noon, she suddenly awoke from her long trance, and stated that "God had told her that she could sit up at the table and eat anything she wanted." At her urgent request her father lifted her from the bed where she had lain helpless so many weeks, and placed her in a chair at the table, when she helped herself to potatoes, bread and butter, fried eggs, and pie; everything but meat she partook of freely, and when her mother, frightened at the quantity she was eating, chided her, saying she feared the consequences, the girl stated that God had told her nothing she ate would hurt her, and that she must cheer them (her parents) up. They had been low-spirited for many weeks, and now must be encouraged. There had been evil spirits in possession of her, she said, for ten weeks, but the good spirits had overcome them. She stated that the doctor was giving her just the medicine she needed, with this exception, that since he had seen her she had a soreness in her throat and lungs; and she wished them to write to him for a prescription to remove that difficulty. Upon being laid back in the bed, her voice changed and she spoke as follows:

"I am Julia's aunt (naming an aunt who had died in Germany). You have had a very sick girl. She is not now exactly sick, but will show a change in a few days. If God should take her away, you must not mourn for her, for she is His child. I think, however, that she will be permitted to remain with you."

She also told several other singular things, which we are requested to withhold at present. On Wednesday afternoon, when we last saw Julia, she was in that same death-like stupor that has characterized her disease for nearly eleven long weeks. Her body is wasted to a mere skeleton. Altogether it is one of the most singular cases of which there seems to be any record.

WITCHCRAFT AND SOMNAMBULISM.

To the Editors of the Enquirer:

In your issue of yesterday I noticed, on the third page of your interesting paper, an article describing an alleged case of witchcraft, etc., of which, if you will allow me, I will give the explanation. I have made the subject under consideration and its connections a specialty, and have studied it for the last thirty-five years, and was in hopes that the ignorance of the age of Cotton Mather had passed away forever, but I am sorry to see that so few still understand the nature of the case.

To be perfectly understood, it will be necessary to make a few preparatory remarks, and in doing so will be obliged to condense matters by simply giving the facts without entering into lengthy explanations at this time.

The case of the girl above alluded to is simply one of natural somnambulism, in which the subject is unconscious of her powers, and consequently swayed by what she has been made to believe before she fell into that condition. All physicians, who have made the subject of somnambulism their study, know full well that catalepsy, clairvoyance, and her speaking in a language foreign to her previous waking moments, etc., are powers, symptoms, and acquirements natural and possible to persons while in that condition.

In the artificial somnambulic state—or, in what has been called the Mesmeric or animal magnetic condition, precisely the same powers, sympathies, etc., are possible—and the only difference between the two states is that, in the latter, they can be taught to make themselves rigid, or cataleptic, clairvoyant, etc., at will, and to awaken or fall into the state at pleasure. These facts, I am aware, are not generally known; but there is no excuse for them being so, as an unprejudiced investigation would soon prove, what I have stated, to be true. Permit me, in explanation, further to state, that the girl had an attack of rheumatism—a very natural result of taking cold, and which generally arises from imprudence or a transgression of the laws of health, without any reference to imaginary causes—of which witchcraft is the most impossible. Her falling into what was supposed to be a trance condition, to my mind, was clearly only a natural somnambulic condition (into which thousands of other persons have fallen—and are daily doing so now—without any reference being had to witchcraft as the cause). The case of the girl is only an extreme of the same condition, and is always the result of natural causes, while the *seeming* convulsions, which were exhibited, were occasioned by the girl's imagination—having been taught, and believing, that her grandmother was really and truly what had been represented to her, viz.: a witch.

The High and Low German spoken by her "in her lucid intervals"—and which was foreign to her before she entered that condition—was acquired during the time she was in that state; and many similar cases have been recorded in medical works, and are perfectly familiar to those who understand the true nature of the condition, and the powers of persons while in it.

The circumstance she related in regard to some ladies of South Bend was simply a case of clairvoyance, and similar feats of seeing are accomplished every day by hundreds of other clairvoyants, and is the result of natural powers possessed by all who can get the eye into a somnambulic condition.

Believing that she was under the influence of her grandmother caused all the difficulty and her misery, and if she had known her powers while in the condition she could have thrown herself out of it or fallen into it at pleasure, or could have done either at any moment she decided upon.

What she said in regard to God, evil and good spirits, and the medicines she needed, etc., was real to her, but the cause was certainly the result of her mind, and not understanding the nature of her condition, she was, of course, influenced by circumstances, ignorance, and her surroundings.

In explanation of what was spoken through her, in the last paragraph, by her deceased aunt, I have but to say that she was then, also, in a somnambulic state, and that if disembodied spirits can speak through persons at all, it can only be done while they are in that condition.

This, however, is a matter for all to decide for themselves. I can here only state facts, and hope that this interesting subject will receive more attention from scientific minds than has heretofore been bestowed upon it. Respectfully,

WM. B. FAHNESTOCK, M.D.,
Of Lancaster, Penna.

CHAPTER XV.

OF INTUITION.

INTUITION is supposed, by some, to be a faculty possessed by somnambulists, and is described in the "Practical Manual of Animal Magnetism," by Dr. A. Teste, as "a sublime instinct which all at once initiates the individuals, in whom it is developed, into the most obscure mysteries of his intimate nature."

"It never," he continues, "could be imagined with what tact, accuracy and precision, somnambulists account for anything which takes place within them. They are literally present at the performance of all their organic functions; they detect in them the slightest disorder, the minutest change. There are no affections so slight or so latent, even those which, at the commencement of their existence, not only do not give occasion to any external symptoms, but do not betray themselves by any species of internal suffering; there are none, I say, which escape the investigation of the somnambulist. Then of all this he forms a clear, exact, and mathematical idea. He could tell, for instance, how many drops of blood there are in his heart; he knows, almost to a gramme, how much bread it would require to satisfy his appetite at the moment, how many drops of water would be necessary to satisfy his thirst, and his valuations are inconceivably exact. Time, space, forces of all kinds, the resistance and weight of objects, his thoughts, or rather his instinct measures, he calculates, appreciates all these matters by a single glance of the eye."

I have made the above extract, in the first place, to show the extent of knowledge upon this subject in France at the present time; and, in the second place, as an apology for those who nave heretofore believed in the science. This is but a specimen, however, of the hundred thousand absurdities which have been ascribed to somnambulists, and it is no wonder that the intelligent part of the community should have doubted the whole, when such palpable errors have been proclaimed and upheld by the professed leaders of the science.

When somnambulists enter the state under the *belief* that they have such extraordinary powers as Dr. Teste has ascribed to them, whatever they may say or predict *shall happen to themselves*, will certainly take place; but what they say or predict of any other person will generally be incorrect, particularly if that person has not been made acquainted with what has been said.

This proves that what they determine upon is the cause of the effect upon themselves, and consequently, is a result of a predetermination on their part that it shall be so.

This is the reason why Dr. Teste's patients told

correctly what he stated; but this correctness will only hold good when they speak of matters which are connected with themselves.

In concluding the same article upon intuition the doctor himself, is inclined to doubt the truth of his assertions, and, although he does not altogether recant what he has said, he attempts, by calling clairvoyance to the aid of intuition, to explain the difficulty, and says that:

"Though the phenomena of vision contribute much in lucid somnambulists to perfect and probably to rectify intuition, it is still beyond all doubt that this latter is not necessarily subordinate to a perfect clairvoyance. I admit, however, that the integrity of the one of two faculties seems to me a guarantee for the excellence of the other; and as we possess no means of verifying most of the phenomena of intuition, I do not give implicit credence to these phenomena, except in the case of somnambulists endowed with clairvoyance.

"Intuition is to clairvoyance what physiology is to anatomy, with this difference, however, that somnambulists, even the least lucid, *feel a priori* the vital actions which are taking place within them, and have no occasion for organographic notions in order to be endowed with a very passible physiological appreciation."

It is very evident, to my mind, that intuition plays no part in obtaining correct information of any kind, and all that is or can be obtained by somnambulists or persons in any other state, is through *the senses* or *faculties* properly so called.

I deny that it is possible for the best somnambulists to do what has been above ascribed to them. They can estimate or guess as well as when awake; but if they are not very careful, may be as often wrong as they would be in a natural state. I admit that all their faculties, while in this state, are capable of becoming more active than when awake; and on that account they may often have, perhaps, estimated better than they could have done when in a natural state; but I deny that they have or can gain any correct knowledge by what is called intuition or instinct. The words intuition and instinct are understood to mean "Knowledge not obtained by deduction of reason, or an internal impulse to act in a certain way in ignorance of the cause."

There is a certain craving or a propensity in each of the faculties, produced by the activity of the functions (love) in each of the organs, and which activity is, in my opinion, always brought about by the outward presentation of something through one or more of the senses, or by a corresponding idea in the mind producing the necessary stimulus; but I cannot conceive that this is even intuition or instinctive, but, in every case, a produced inclination or propensity to act in a certain way according to the nature of the faculty acting.

New born children cry and suck, etc., but it is argued that this is not an act of the understanding, and that it is produced by some internal impulse or instinct.

If there be no *positive understanding or reflection* in these cases it is certainly very evident that children both cry and suck from *certain motives*. When they cry they must have experienced something unpleasant or painful, through one or more of the senses, etc., and when they suck the organ of alimentiveness is active, and has been excited either by the sight, the smell, or some gastric necessity stimulating the organ, and therefore cannot be instinctive, or an effect without a cause in either case.

Animals do not act from instinct any more than men. They possess some faculties common to man. Man has the advantage of numbers, and kinds, but not always of degree.

CHAPTER XVI.

PRESENTIMENT OR FOREKNOWLEDGE.

FOREKNOWLEDGE, properly so called, is a knowledge of something which has not happened, or is yet to come; and of which no person has any previous thought, intimation, fear, suspicion, or knowledge of any kind.

Many cases are related which are said to have been of this description; but I have as yet not been able to see an instance sufficiently clear and unobjectionable to change my belief in its impossibility. The following case has been detailed as one of this character, in a work on "Pathetism," by La Roy Sunderland.

"An intelligent gentleman of this city, and one, by the way, as little given to belief in dreams as any one I ever knew, gave me the following account. Business rendered it necessary for him to visit Albany. But for some reasons, to him altogether unaccountable, he felt very much disinclined to go. After vascillating for some time, he finally took one of the evening boats, and at the usual hour retired in his berth. He soon fell asleep and dreamed that he saw his wife, pale and scarcely alive from loss of blood. She was so near dead that she had become quite cold, and he built a fire, that by warming the body he might bring her to life again. The sight so disturbed him that it awoke him, but on falling again to sleep, he had precisely the same dream again. The cause of his wife's having bled to death he did not perceive, but it now appeared that she was cold and dead from loss of blood.

"On waking in the morning, he found it impossible to banish the impression which these dreams had made upon his mind; and soon after reaching Albany he received a letter summoning him to return, and stating that his wife, within an hour after his departure, had met with the accident from which she had well nigh flooded to death.

"On returning, he found that she had, indeed, been so near dying that the physician and friends had given her up, supposing that nothing could by any means save her life.

"When he left her, he had not anticipated anything of the kind, and is not conscious that such an accident ever entered his mind. And I should add that this gentleman has no faith in clairvoyance."

This case I consider to be purely of a clairvoyant character, and although the gentleman to whom it happened may not believe in such a power, I am persuaded that his information could not have been received in any other manner.

The gentleman's mind, in the first place, was much disturbed, and, although he could assign no reason for it, it does not follow that because he felt so, that something of an unpleasant nature must necessarily happen.

Such feelings or presentiments, *the cause of which I have already explained in a previous chapter*, very often steal over the minds of most persons when they are about to leave home, particularly when they are not very anxious to go; because we know that there is a possibility that something may happen; but the functions of the brain, although we may not be conscious of their doing so, are almost always on the alert, and reaching out often observe things in the outer world that are taking or about to take place, and produce presentiments which we may not always be able to define. This was no doubt the case with the gentleman in question, and was the cause of his uneasiness, and his vision which followed I think can be explained or accounted for in the following manner:

He felt uneasy and disturbed about leaving home, consequently his thoughts were upon it. He went to bed, fell into a somnambulic condition, and his mind naturally wandering home, he saw his wife in New York, precisely as she lay, pale and scarcely alive from loss of blood. What followed was very natural, and his finding his supposed dream true only proves that he saw correctly, and any other good clairvoyant could have done the same thing for him, had their mind or attention been directed to his wife at the same time.

The following, which is also detailed in the same work and ascribed to sympathy by the author, was evidently the result of accident, or at most the effect of premonition or presentiment.

"Judge Stelling, in his almanac of 1808, relates a remarkable presentiment of a minister, who was taking a walk with the intention of visiting a rocky mountain near his house, and of enjoying the beautiful view from it. While approaching the summit of the mountain he felt restless and uneasy; unable to explain this feeling, he asked himself whether it was right for him to spend his time thus idly; and busied in such thoughts he stepped aside for a moment to seek a cool place under a wall formed by the rock. He had scarcely left the narrow path leading to the top of the mountain, when a large stone, breaking loose from the rest of the rock, with great vehemence struck the spot where, one moment before, he was standing."

Many similar instances of presentiment have been related by various authors, but they can all be accounted for upon reasonable principles. Many more could be stated where nothing has happened after similar feelings have been experienced. In these cases the functions of the imagination in the faculties *reaching out* and acting, have created ideas which produce the feelings experienced, and are as real where nothing happens as where something does take place, a knowledge of which was before obtained by the reaching out of certain functions, independent of the subject's consciousness.

CHAPTER XVII.

I.—OF INTERIOR PREVISION.

THIS is considered a species of fore-knowledge, and the following description of it by Dr. Teste will give the reader an idea of what is meant by the term.

"Lucid somnambulists have not only the consciousness of their present philosophical or pathological state, but they can even announce by a sort of fore-knowledge which belongs only to them, all the modifications destined to occur in their system. It is a calculation on their part, a process of deduction, by means of which they determine from that which exists that which does not yet exist.

"This is what no one can tell; but what is quite certain is, that the instances of interior prevision, that is, of prevision bearing on occurrences and events relating to the individual himself, are exceedingly numerous. Every author on the subject cites several of them.

"Heter Cazot, a somnambulist mentioned in the Report of 1832, predicted several weeks beforehand the day and hour when he will be attacked with a fit of epilepsy. He not only determines the instant when the fit is to come on, but also its violence and duration. The commission which verifies the fact vouches for its authenticity.

"Paul Villegrand, who is also mentioned in M. Husson's work, predicted a long time beforehand all the phases of his disease, and everything occurred precisely as he predicted."

There is no doubt that these things have been done, but I differ materially in my opinion as to the cause. It is well known to me, that when a resolution is taken, a belief cherished, or a determination formed by persons while in this state, that, when they awake, although they may know nothing about it or relative to it, they always do what has been so resolved or determined upon at the time appointed or specified.

I have had a lady, who, for the last ten years, had not tasted coffee, and who always disliked anything sweet, to ask for a cup of sweetened coffee at a certain time, although she could not at the time tell the reason why she did a thing so contrary to her habits. Ten days before, she had agreed, while asleep, to try whether an experiment of the kind could be effected in her case without her knowing anything about it while awake; and when the time came, she declared that she felt almost an irresistible propensity to drink a cup of sweetened coffee, and did so, with a perfect relish. This lady could create or relieve pain in any part of her body, or even forget her own name when she awoke, if she made the resolution to do so while in this state.

I could relate many other cases of the kind were it necessary; and I have stated the above to show that it is possible for things of the kind to take place independent of the subject's knowledge when awake, although they may have doubted the fact before they tested it themselves.

This is precisely the case with those who predict or say that certain things shall come to pass, or take place at a certain time; and is the reason why all predictions relating to themselves are always verified.

It is, therefore, *highly important, in all cases, not to let them determine upon such events, unless* they are of such *a nature as to be beneficial to them in their waking moments.* Too much stress cannot be laid upon this important point, nor too much caution given to those who entrust themselves into incompetent hands, as improper and highly injurious consequences may follow for life if such persons do not understand the true nature of the various phenomena which are peculiar to this state.

This strange power, or faculty, is one of the most important, however, in the catalogue of phenomena belonging to this state; and I have taken advantage of it, as I have before stated, for the purpose of curing diseases, removing unpleasant habits, or unfortunate affections, etc., and of which I shall speak more fully when I treat of these subjects.

II.—OF EXTERIOR PREVISION.

External prevision is said, by Dr. Teste, to be the

"incomprehensible faculty of predicting during their somnambulism events with which their existence will be mixed up, but the cause of which is evidently foreign to their system, or one to which their system bears no relation, and cannot have any species of connection with it that can be at all explained."

Many cases of this kind of prevision are said to have taken place. That of Madame Hortense, related in Dr. Teste's "Practical Manual," I here give a condensed account:

Madame Hortense, it appears, had been in a state of Artificial Somnambulism frequently before the present sitting, in which she predicted, among other things, that she *should take a fright at something on a certain day*, which she mentioned, the result of which would be a *fall* and a *miscarriage*. Dr. Teste endeavored, by questioning her, to ascertain where and at what she was to become frightened; but she could tell nothing about it, and insisted that nothing could prevent it. To his question—"Do you know precisely what you are to suffer?" She replied, "Certainly; and I am now going to tell you. Tuesday, at half-past three, immediately after having had a fright, I shall have a weakness come over me, which will last eight minutes; after this, I shall be seized with very violent pains in the kidneys, which will last the remainder of the day, and will continue for the entire night. On Wednesday morning I shall commence to lose blood; this loss of blood will increase rapidly, and become very profuse; however, there will be no occasion to be alarmed at all this, for it will not compromise my life. Thursday morning I shall be much better; I shall even be able to leave my bed for almost the entire day; but in the evening, at half-past five, I shall suffer another attack of hemorrhage, which will be followed by delirium. The nights, Thursday and Friday, will be favorable ones; but Friday night I shall lose my reason." Her husband, who was present, very much moved at what she had stated, asked her whether she would be long delirious? She replied, "Three days;" and added: "Be not alarmed, Alfred, I shall neither remain bereft of reason, nor shall I die; I shall suffer, that is all."

She was aroused, and remembered nothing that she had predicted or had passed in her sleep.

The fatal day arrived, and Madame Hortense was persuaded to enter the sleep again, and again predicted that *"Between three and four o'clock I shall take a fright at something; I shall have a fall; the result will be an abundant hemorrhage."* To all the doctor's questions respecting the cause and where it was to be, etc., he could obtain no satisfaction.

She was awakened, and again knew nothing of what had transpired. She was watched through the whole day by her husband and Doctor Teste. Every precaution was taken by them to prevent her being frightened. She noticed these unusual attentions, and, of course, wondered at them; but they evaded her questions as to the cause, and told her nothing about her predictions. She eventually requested that they would permit her to withdraw, and she did so in company with her husband.

They had left the room but a moment when the doctor heard a piercing shriek and the noise of a body which had fallen. The following is the doctor's account of what had happened: "I ran up; at the entrance of the water-closet Monsieur * * * holds his wife, with all the appearance of dying, in his arms. It is she, then, that cried out, and the noise heard was occasioned by her fall. The moment she let go her husband's arm to enter the water-closet, a rat, of which she had an incredible horror, in a place where one had not been seen for twenty years before, presented itself to her view, and gave her so great and so sudden a fright that she fell back, without there being a possibility of sustaining her."

All the rest is said to have taken place precisely as she had predicted.

The above, at first sight, appears inexplicable; but when we examine the facts collectively, and reflect that it would have been as easy for her to have told that *the rat was to produce her fright* (granting that she had this power at all) as it was for her to tell or predict that she was to have a fright at a certain time, and that it was to produce the train of consequences which she so minutely described, and her not being able to predict this, nor the place where it was to happen, etc., proves to my mind that persons in this state have no power to foretell anything correctly that does not depend upon themselves.

This, then, is nothing more than that which was described as being interior prevision; and the same explanation which was given of that is applicable to this also; and if the rat had not frightened her by the merest chance in the world, something else would; for frightened she would have been at the predicted time, if nothing more frightful had presented itself than her own shadow. What followed was upon the principle, that the prediction is the cause of the effect; and if she had not been suffered to predict, or had previously been made acquainted with the true nature of the sleep, and her absolute powers therein, that the above predictions never would have been made, and, consequently, what followed never would have taken place.

It therefore shows the extreme folly, if not criminality, of encouraging or permitting subjects to predict or make resolutions which may be inimical, disastrous, or contrary to their well-being, whether they are asleep or awake.

III.—PROPHETIC DREAMS.

Prophetic dreams are referable to the same cause, viz.: Somnambulism, and can all be explained upon the principle of clairvoyance.

IV.—WITCHCRAFT.

Witchcraft has nothing to do with Somnambulism, and is only an effect produced upon the minds of those who have been made to believe that certain persons were endowed with certain powers.

If persons ignorant of these facts are made to believe that a certain person, styled a witch or a wizard, has power to produce certain effects, such effects will, sooner or later, follow that belief; not because such a person has such powers, but because the subjects believe that they have. The maladies produced in such cases assume various shapes, and are referable to the impressions originally received, or the superstition and whims of the subjects themselves. For a case in point, see page 186 of this work, viz.: on Witchcraft and Somnambulism, as published in the *Daily Enquirer* of Cincinnati, May 15th, 1869, under the head of "TRANCE."

CHAPTER XVIII.

SYMPATHY.

NOTWITHSTANDING all that has been said in favor of sympathy, and all the arguments used and cases quoted to prove that such a thing exists between bodies entirely distinct from each other, and, independent of direct communication of any kind, or that it is the means through or by which all the phenomena are exhibited in cases where one part of the body or different bodies, etc., not originally affected, shall be affected by another originally so, I cannot subscribe to.

The phenomena ascribed to it are exceedingly numerous; and it has almost become, like electricity, the cause of everything that cannot be explained by an imperfect philosophy.

That there is a correspondence, or a reciprocal adaptation of one part of the body to another, and of the whole body to the mind, is clear to my reason, and the effects which one produces upon the other is often remarkable; but I deny that they are effected through sympathy, or anything like what is understood by that term.

In a well-disposed person we see a remarkable coincidence in the shape of the head, the qualities of the mind, and the expression of the countenance, etc., but is this concurrence the effect of sympathy? Who would be willing to subscribe to such a doctrine? Is it not more rational to ascribe the appearance of the face to the positive qualities of the mind producing the effect directly through the nervous system, without a reference to anything else, or seeking for something which of necessity must render the philosophy of the matter less clear?

If we titillate the fauces, vomiting will ensue; or, if we see food, the salivary glands will be excited. Disgust, swinging, sailing, or a blow upon the head sometimes produce vomiting. Diseases of the thigh or hip-joint are sometimes felt in the knee; those of the arm, in the elbow; and those of the liver, in the right shoulder, etc.; but all these facts do not prove that they are caused by sympathy, and not by a direct communication, first with the brain, and then to the part directly or indirectly affected.

We might as well ascribe the vibration of chords to the same thing, (sympathy,) or say that, when a finger is pricked, the brain receives the impression by sympathy.

When the fauces are titillated, the nervous system conveys the feeling to the brain, and the mind, through the same system, affects the stomach, and produces vomiting.

When we see food, an impression is conveyed to the mind through or by the faculty of seeing, and the mind, exciting the glands through the nervous system, produces a flow of saliva, etc. Disgust, swinging, sailing, or a blow upon the head produce vomiting because the brain, or a certain portion of it, has been disagreeably affected, and the impression has been conveyed to the stomach by direct communication through the nervous system.

Several cases are recorded where similar affections are said to have taken place at the same time, in twins, who were at a distance from each other when such affections were simultaneous.

I quote the following as a supposed case of double sympathy, which was communicated to the Royal Academy of Medicine, by M. Cagentre, in 1824:

"Twin brothers were affected precisely alike for a number of years. Whatever indisposition one suffered was suffered by the other at the same time. Derangement of the alimentary canal, intestinal worms, etc., always made their appearance in both at one and the same *moment*, and the symptoms in both were of equal intensity. Dr. Nourel carefully watched them after their return from a fifteen months' stay in the country, and verified the observations of the nurses. He found that quotidian intermittent fever commenced and terminated on the same day in both; both had acute conjunctivitis together, and also colic, which lasted in each for twenty-four hours. Two molar teeth made their appearance in each at the same time. These things took place in 1831. In 1832 they had different eruptions, but both suffered contemporaneously. In the winter they had bronchitis together. In 1833 they were attacked with measles and after these with scarlet fever; in these diseases each twin had symptoms precisely similar to the other, and the commencement and termination in both were precisely at the same period. In 1834 they had each ear-ache and intermittent fever together; and also vesicular eruption on the back of the neck. But their *dispositions* were entirely opposite; one was thin and lively, the other robust and indolent."

In the above case of the twins, I have but to repeat what I have so often stated, that the cause of all the above affections was produced by the reaching out of the faculties of the boys, independent of their knowledge, and clairvoyantly seeing each other's condition, thus producing, upon the principle of seeing disease in others, like diseases in themselves.

All physicians are familiar with the fact that seeing even with the natural eye disease in one person often produces like affections in others. Convulsions, religious "loss of strength," cholera, and even small pox, have been produced in others, with no other communication than by the sight.

In this case an impression is conveyed to the brain which, being of a disagreeable nature, produces fear or a *belief* that like effects will follow, thus causing the very disease that they desired to shun.

When the religious faculties are excited and act in unison with the marvellous, the cautious, and the imitative faculties, singular affections are often produced, of which the following account by Dr. F. Robinson, of Tennessee, is a remarkable instance.

"The churches," says the Doctor, "in these States at that period were small and uncomfortable, and the people from necessity assembled in the open field at extraordinary meetings. These meetings lasted from three to five days. They remained upon the spot day and night, and worshipped their Maker incessantly. The outward expression of their worship consisted chiefly in alternate crying, laughing, singing, and shouting; and at the same time performing the greatest variety of gesticulation which the muscular system is capable of producing. It was under these circumstances that some of them found themselves unable, by voluntary efforts, to suppress the contraction of their muscles; and, to their own astonishment, and the diversion of many of the spectators, they continued to act from necessity the curious character which they had commenced from choice.

"The disease no sooner appeared than it spread with rapidity through the medium of imitation.

"Thus it was not uncommon for an affected person to communicate it to a greater part of a crowd, who from curiosity, or other motives, had collected around him.

"The contractions are sudden and violent, such as are denominated convulsions; being sometimes so powerful when in the muscles of the back, that the patient is thrown on the ground, where for some time his motions more resemble those of a live fish, when thrown on land, than anything else to which I can compare them. During the intermission a paroxysm is often excited at the sight of a person affected, but more frequently by the common salute of shaking hands.

"The sensations of the patient in a paroxysm are generally agreeable, which the enthusiastic class often endeavor to express by laughing, shouting, dancing, etc. Fatigue is almost always complained of after violent paroxysms, and sometimes a general soreness is experienced. It has not proved mortal in a single instance within my knowledge, but becomes lighter by degrees, and finally disappears."

The "Jerks," the "Rolling exercise," the "Barks," etc., are of the same character, and are produced by the same cause.

I.—CLAIRVOYANCE.

When any of the senses are in the somnambulic condition, they become what I call, for want of a better term, clearminded. Clairvoyance relates to the eye only, and is *internal perception, or perception without the aid of the external eye.* From what has already been said in this work respecting clairvoyance, the reader must, in a great measure, be acquainted with my views, but it is still necessary to give some illustrations in detail, and other explanations connected with this subject.

With this view I give the following letter which I sent to the editor of the "Magnet," and published in vol. II., No. 1, of that work, for June, 1843. I now give the terms as orignally sent. They were altered to suit the views of the editor, which I cannot subscribe to. The letter is as follows:

"DEAR SIR:—I have been engaged in attending to persons while in a state of Artificial Somnambulism, principally for the cure of diseases; yet, whenever I have found the sight of such persons to be clairvoyant, I have made it a point during the time necessary for them to remain in it, to study this remarkable phenomenon; and it has always given me pleasure, on similar occasions, to exhibit its peculiarities to others who were desirous of witnessing it.

"My audiences have, with few exceptions, been of a private nature, and were composed generally of the most obdurate and inveterate skeptics, requiring me to make many and often repeated experiments, with such care and circumspection as their unbelief warranted.

"In this manner and by other experiments which I prosecuted with, if possible, still greater caution, I have been able to gather a sufficient number of facts to overwhelm and convince me that their ability to discern without the aid of their eyes is, indeed, true beyond the possibility of a doubt. I have been extremely guarded in my own experiments, so much so, that I would not permit any one to enter the room after having had them to place various articles in boxes—of which I had no knowledge whatever; yet, after taking each box into my hand promiscuously, and having successively asked questions as to their contents, to my astonishment, I found the answers correct, notwithstanding some of the articles were of such a nature as to make guessing correctly out of question. One of the boxes contained a piece of cabbage leaf, 'broken off and not cut,' as Mr. C. S——, the subject, most accurately described it, together with its shape, size, color, uses, and his own dislike to its taste, when prepared for the table.

" When a sense is in the *sleeping condition* of Artificial Somnambulism it is in a perfect state of rest, and lays dormant until it is aroused by the person in whose care they are, or by an effort of their own will which is, at all times, independent of the so-called operator. He may request them to look or to hear, but they can comply or not as they please; and when their inclination to hear, etc., ceases, such senses relapse into their former inactivity, and are then, so long as it is their will they shall remain so, addressed by others in vain.

" I have also made many experiments lately, to test their powers of hearing at a distance, which I have found as extraordinary as their clairvoyance, and as astounding as it is true.

"The phenomena of clairvoyance seems to create more doubt in the mind of skeptics than any other peculiarity which is exhibited by subjects in this state; yet it is, by no means, more extraordinary than that they can, whenever they please, know the will of the operator, experience no pain while under the severest surgical operation, nor hear any noise unless directed to it by him, or is a special desire of their own.

"Many admit the truth of the sleep and all the other phenomena, but cannot believe that they see without the aid of their eyes. This, they say, is unnatural, and contrary to all their preconceived notions, and, therefore, cannot exist, thus building up a standpoint at whose shrine they would constrain truth to worship.

"There seems to be a morbid propensity in some persons to doubt and to proclaim everything false which they cannot understand, or which does not accord with their peculiar views, as if, without any investigation, they had imbibed with their mother's milk universal knowledge, and despised all that did not originate with themselves.

"But I have lately seen many such men bow to the force of truth in spite of *their philosophy*, and would advise others of a similar character, first to assure themselves that they have examined the subject thoroughly, and under such circumstances as to leave no doubt upon their minds before they arrive at conclusions which may be unjust.

"The reign of theory is past, and unless they can produce facts to substantiate their views they will find that their opinions weigh but little when opposed to truth. If they have not experimented personally, honestly, and with a desire for the truth, they are not and cannot be capable of judging.

"Some, indeed, of the above stamp have, after a

brief and unfair examination, left such exhibitions wilfully dissatisfied, and seemed to glory in their ability to evade the truth, verifying the saying that: 'There are none so blind as those who will not see.' But it is a matter of no importance to the science, or to the world at large, whether such men believe or not, as their opinions—no matter what their stations in society may be—will not alter the facts, nor change the nature of that which really exists.

"Facts, stubborn and incontrovertible facts, have settled this question in my mind, and it is with such arguments only, that I have constrained others to believe that the sense of seeing when in this state, enables the subject to discern, not only articles held in the closed hands of others, but objects and scenes at a distance.

"Darkness, matter, and space, seem to offer no obstruction to their view, and I have had them, times without number, correctly to describe and name articles held in the closed hands of others, of which I had no knowledge whatever.

"In the same manner they have described pictures, etc., held behind them, and named persons outside of the house, although their presence was not expected, and they arrived after the subject had been in this state for some hours. They have told the contents of closed boxes at a distance, which they never saw, and named the amount of money, and kind of coin in pocket-books and purses which were held in the hands of inveterate skeptics.

"They have found persons at a distant city, with whom they were acquainted, without ever having been there themselves, and told accurately—neither more nor less—what they had been doing at a certain time and place.

"They have described places and scenes at a distance, where they had never been, to the perfect satisfaction of hundreds of skeptics who, at different periods, requested them to go with them in thought; yet, notwithstanding all these facts, some still hold out and battle in the dark.

"Clairvoyance must be seen to be believed, and I do not censure those who disbelieve it, that have not had an opportunity of witnessing it in its purity; but I do those who, when present on an occasion of the kind, arrogantly assume the umpire and question the integrity of every one, at the same time that they, themselves, use all the unfair means and trickery in their power to thwart the experiments.

"In conclusion, I will state, for the benefit of those who are anxious to try experiments successfully, that it is necessary to put all questions distinctly, and to await their answers patiently, as hurrying them from one thing to another distresses them, and frequently makes them unwilling to answer at all. When I wish to direct their minds to anything, I usually do so by telling them where it is placed or held, and then ask them whether they see it? If they answer in the affirmative, I then request them to name or describe it. This is not always necessary, particularly after the first sitting. The articles, etc., to be described, ought not to be known by the operator before the experiment, nor should there be a word said in the room by any person, respecting them, either to instruct or mislead, as I have often known such interference to thwart experiments which, otherwise, would have been satisfactory."

The views in the above letter, which were published twenty-six years ago, I still hold to be true, and many similar facts have been added to the list, all of which confirm the knowledge previously obtained. But, as I have before stated, it must be remembered that I do not place implicit confidence in the clairvoyance of *all* subjects, particularly when they imagine or are unwilling to look, knowing that they can see what they imagine as well as they can that which really exists, if their minds are not carefully directed.

CLAIRVOYANCE AT A DISTANCE.
Case 1.

Subject, Miss ———. She was requested to observe and to state what Mr. K—— was doing in the next room, *the door being closed*, and the back of her chair toward the room in which Mr. K—— then was.

Answer.—" He is standing in the center of the room, and is holding a chair above his head."

The door was thrown open, and Mr. K—— stood where she said, holding a chair above his head. It will be necessary to state that Mr. K—— was extremely skeptical, and was, of course, not satisfied with one experiment. Several other skeptics were also present, who took great care that everything was done to their satisfaction.

The door was again closed by them, and she was again asked what Mr. K. was doing.

Answer.—"He is standing up, and is holding a pillow upon his head."

The door was again thrown open, and he was found to be standing, holding a pillow upon his head.

The door being again closed, she was once more desired to state what Mr. K. was doing.

Answer.—"He is lying down full length upon the floor."

Her answer was again correct: he was found stretched upon the floor when the door opened.

On another occasion the same subject was requested by several other skeptics to tell what Mr. S. was doing in the next room.

Answer.—"He is standing up, and is holding the *piano-stool* upon his *right shoulder.*"

Her answer was correct; and in like manner she told that he was holding a *note book upon his head;* and again that he had thrown *a shawl about his shoulders,* and *had placed a bonnet on his head.*

The same precautions were taken by the gentlemen to prevent deception that had been used on a former occasion. The door was guarded closely, and opened by themselves, and the positions which Mr. S. assumed were not premeditated by him, but assumed upon the instant after the door had been closed. Deception was therefore out of the question.

Case 2.

Subject, Mrs. D. She had been afflicted with dyspepsia and nervous headache for several years, and had entered the state twice before under my care, and on the present occasion was at a neighbor's house, about half a mile from her home. She came over in the afternoon for the purpose of entering the state, and was to remain there while I tarried in the neighborhood. In the evening, after having performed various experiments in clairvoyance, her husband came over to see us, and as he had no faith in her ability to "*see with her eyes shut*," he requested that if she was able she should look home and see whether everything in a certain room was as she had left it.

After she had looked, she remarked that he must have given himself a great deal of trouble to strip the children's bed, and to disarrange the furniture. Not satisfied with this, he requested her to say what he had placed upon a certain dresser. She at first seemed very unwilling to look, but at last did so, and immediately said that she knew what it was, and desired to know what possessed him to place the *small chest* up there. Her husband then stated that before he left home, he had stripped the children's bed and disarranged the furniture in the room, and placed the *small chest* upon the dresser.

Case 3.

Subject, Miss H——. This young lady had been afflicted with epileptic convulsions from her childhood, and was now on a visit to Lancaster for the purpose of trying what effect Somnambulism would have upon her disease. She had been in the condition several times, and since her first sitting has had no return of her fits. Upon this occasion, after entering the state, she was requested to throw her mind home—about four miles distant—and see what was going on there. After stating many things respecting the family, she said that they were hitching up the horse into the *small wagon*, and that her mother was getting ready to visit Lancaster, but wondered why they did not take the buggy; and after a pause said, "They are now getting into the wagon, and are coming towards Lancaster." In about an hour afterwards her mind was again directed to them, and she said they were almost in the city. In about ten minutes after, the vehicle was driven up to the door and her mother entered the room. This visit from her mother was not expected, and the roads being heavy, the *small wagon* was employed instead of the buggy, which was usually made use of for that purpose. This young lady has never had an attack of epilepsy since, was married some years ago, and is raising a fine family.

Case 4.

The following was sent to the editor of the "*Magnet*," and published in that periodical in November, 1843·

Subject, Mrs. H——, of most exemplary character, who has been laboring under a nervous affection of the eyes and lower extremities, *rendering her perfectly blind and lame* for two years, and who has been entirely restored by entering this state.

She was requested while in this condition to tell what a certain gentleman had in his yard attached to his house, at a distance of several miles. When asked whether she would look at it, she replied that she did not care about going there, but would look; and upon doing so, asked me what kind of an animal it was.

I told her I did not know what the gentleman had there, as he was very careful not to tell me. "Well," said she, "I have seen one like it in the museum, but I never saw a live one. *It is a raccoon.* He is fastened to the oven, and is now lying in a box near it asleep."

This was acknowledged by the gentleman to be the fact. He had placed it there *that evening*, and came over immediately afterwards, expressly to test her powers.

Some time previous, the same lady was requested by a skeptic to visit a gentleman's apiary at a distance, and to tell the number of hives and the condition of the bees, which he represented to be flourishing. When asked, she remarked that he had about twenty hives, but that the bees were all dead. This seemed strange to me, and I asked her whether she was sure that they were all dead. She said, "Yes, you will see, they are all dead." The gentleman then stated that such was the truth, and that the fact had not been known to any one but himself.

On another occasion she was requested by a neighbor to visit his house, and to state where his wife was, and what she was doing. After stating several things respecting her, to his satisfaction, she asked me, when Mr. B——'s tree had blown down.

I asked Mr. B—— whether that was the fact, but, instead of answering, requested me to ask her whether it was all blown? She said, "No, about the half of it, and it is lying there still."

Such was the fact; half of the tree—a very large willow which stood before the door—was prostrated by the storm in the night, and next morning early the fact was stated to the owner, *unasked*, and independent of any communication between the houses. His intended visit was unexpected to us, and the distance between the houses is about four miles.

When questioned as to the reason why she noticed the tree, she remarked, that when she came to the front door, she found she could not enter the house without climbing over the fallen tree, and found it necessary to pass around the house to enter the kitchen.

Case 5.

Subject, Miss Z——. Of her own accord entered the condition, for the purpose of visiting an aunt, who lived about fourteen miles distant, and after she had cast her mind to the place, she seemed to be delighted, and when asked why she was so much pleased, she stated that her aunt and her two cousins were making preparations to visit Lancaster.

About two hours after, her sister, with the view of teasing her, remarked: "Ah, Miss, you must have been mistaken about aunt's coming to-day. The cars have arrived, and she has not yet come, although time enough has elapsed for the omnibus to have been here long ago."

"Ah, indeed!" replied Miss Z——. "It is you that are mistaken. They are not coming in the cars. They are in their own carriage, and will be here directly." Soon after the carriage was driven to the door, and her aunt and two cousins stepped into the room.

This visit was entirely unexpected by the family, and when she stated the fact, their coming was doubted, and they could not realize it until the carriage was at the door.

Case 6.

It was agreed, between a gentleman and myself, to test clairvoyance at a distance of sixty miles, and when in Philadelphia, he was to visit a certain house known to me, and there to do certain things which he was to determine upon and note. I, being in Lancaster, was to have one of my subjects, who had never been in Philadelphia, to say what he was doing there, at a certain time.

He departed from the city in the morning train, and in the evening of the same day, Miss Z—— entered this state the twelfth time, and when taken in thought to the appointed place, she declared that he was not there; that the house was closed, and not occupied.

This seemed strange, as it was the time we had set, and I could not think that he had forgotten his engagement, nor could I tell why the house should be closed. Under these circumstances, I was at a loss to know what I should do, and although I had the utmost confidence in her powers, having sufficiently tested them before, I was not yet prepared to believe that she could find him in a city where she had never been herself. But as I could lose nothing but the time spent in the experiment, I desired her to see whether she could find him. After three or four minutes had elapsed, she said that she had found him, and that he was in the third story of a house, in a room alone, containing one bed, several chairs, a bureau, and a wash-stand, etc., and that he was *standing up at a covered* bureau, with a parcel of papers spread before him, and that he was figuring with his pencil. After a few minutes, she remarked that "he is now gathering up his papers; now he is going down stairs; now into the street; and down the street; he is now about to enter a large building; he is speaking to some one at the door; it is Mr. L——. I know him; he is now inside. This must be the theatre," and, as if speaking to Mr. ——, she said: "Take a seat, Mr. ——."

She then described the house, and said it was crowded.

The following is Mr. ——'s account, which I received just after he had stepped out of the cars, where I had gone to meet him, upon his return to Lancaster.

"I arrived in the city of Philadelphia about the usual hour, and while down street that afternoon, attending to some business, I ascertained that the house I intended visiting in the evening, for the purpose of performing my part in the experiment, was closed. I, therefore, of course, could not go there, but went to my boarding-house, and as I thought, that I had lost ten dollars in one of my transactions that afternoon, I retired to my room, in the third story of the house, for the purpose of finding where the mistake lay, and at the time appointed for the experiment, I was standing at a covered bureau, with my papers spread out before me, and figuring with my pencil to find out the error.

"Finding all correct, however, I concluded to go to the theatre, and gathering up my papers, I went there, met and spoke to Mr. L—— at the door, and then entered the theatre, which I found very crowded.

"My chamber contained but one bed, a bureau, a wash-stand, and two or three chairs."

Case 7.

Mr. ——, a gentleman who had frequently witnessed the powers of clairvoyants in seeing things, &c., both in and about the house, became desirous of entering the state himself, for the purpose of testing the power of seeing things at a distance. As he visited the city of Baltimore frequently, he requested certain acquaintances there, who were skeptical, to place something at a particular locality in a certain house, after he had left the city, for the purpose of testing his powers of vision, should he succeed in entering the state when he returned to Lancaster.

This was complied with by his friends several times, but as he could not succeed in entering the state perfectly, after the third trial, he requested a lady, who was accidentally present, and had been in this state repeatedly, to enter the state and to look at it for him, so that he could convince his friends in Baltimore that it could be done.

Subject, Mrs. E——. She had never been in Baltimore in her life, and after she had entered the state, it was necessary—as I was not acquainted with the location of the house—for him to convey her in thought to the appointed place. Having done so, I requested her to describe the room, which she did to his satisfaction, and as the thing to be looked at was to be at or about the time-piece, I directed her attention to it, and desired her to look whether there was anything about the clock which did not belong to it. She said she saw something dark there, which looked like a bottle, but that she felt as if she were going backwards, and could not keep herself there long enough to see it distinctly. This being the case, and finding that her mind was wandering about the city I directed her to look about the city, and after

I had taken her to the Washington Monument and various other places of interest, I desired her to go back to the clock again, and to go up to it, and to take the article which she before described as being a dark bottle, into her hands, and to examine it minutely, so that she could be certain as to what it really was. After having done so, she declared that she now saw it distinctly, and stated that "it was a *dark bottle, about the length of her index finger*, and was suspended by a *white string, tied about its neck, that it was empty, and had no cork.*"

The gentleman left Lancaster for Baltimore the next day, and when he returned he stated that, as he approached the house of his friend, in Baltimore, where the thing to be looked at was to be placed, he saw his friend at the door, and, as he came up to him, his friend immediately asked him to tell what he had seen placed near the clock. After he had related the circumstances and told what the lady said, his friend produced the bottle, which had been suspended at the time agreed upon, and which, to their mutual astonishment, they now saw she had described to the very letter. The gentleman brought the bottle with him to Lancaster, with a piece of the white string still attached, and after it was shown to Mrs. E——, she declared that it was the very same which she had seen suspended in Baltimore.

The bottle is of a very dark brown color, and looks nearly black when not held up between the light and the eye, of a peculiar shape, and not easily mistaken.

It is about the length of an index finger, and was empty, and without a cork or stopper.

A purer case of actual clairvoyance could not be desired, because there was no person in the room that knew or had any suspicion of what might be placed there. It was to be placed there for a certain time and then removed. We were all ignorant of the fact, and could not tell or say whether there was anything there or not; and it was not known to any of us, whether she had seen correctly until the gentleman returned with the bottle, which she delared was the identical one which she saw there suspended, although she had never been in the city herself, and neither of us had any knowledge of the locality of the house, etc., but the gentleman who desired the experiment, and the article was not placed there until he was in Lancaster.

I could relate many other cases of clairvoyance at various distances were it necessary; but if what I have already related be not credited, more will not.

Some persons, who have had no practical experience, and profess to be skeptical, have intimated that I have been deceived, and that what I accept as facts are the results of my imagination. I ask for the proof! Let them state in what instance I have been deceived, or contradict what I have stated in regard to any of the phenomena which I consider to be truths. If they cannot do this their assertions are unjust, and the weakness ascribed to me must fall back upon themselves. But I have not been deceived, and am confident that if the most skeptical person had observed the proofs that I have witnessed they would also be satisfied of the facts, although they might not be able to explain the why or the wherefore. Let them, therefore, not say what they are not prepared to prove. What I have said I have proved, and can do so again.

Much has been said, by various authors, about the powers of clairvoyants in knowing or seeing what has passed or is yet to come. I have already given my views upon these points, and will here but remark, that when a circumstance is once passed, it is lost to them forever. It is true, they may get such knowledge from others who were present at the time or knew about it, or they may speak of what has passed from a previous knowledge of their own, but they cannot see it in any case independent of the imagination, any more than they can unerringly foresee that which is to come.

The following case is quoted from the "*Magnet*," March number, 1844, and is supposed, by some, to be an instance of seeing the past:

"Some time during the month of January last, Mrs. ——, of the village of A. A., in the State of Michigan, missed from her parlor table a beautiful little gold watch. It was taken one evening, while no member of the family was in the room. The whole affair was enveloped in mystery. Suspicion rested on no one in particular, in the mind of Mrs. S. or her husband. Careful search and inquiry were made for several weeks, but all to no purpose. The singular disappearance of the watch remained an inexplicable secret, locked up in the bosom of the unhappy young man who had ventured to commit the deed. A few months passed away, and the matter was nearly forgotten.

"In the spring—in the month of April, I believe—Mr. D. B., the distinguished scholar in the science of Animal Magnetism, visited A. A. for the purpose of lecturing and exhibiting facts and experiments in proof of the pretensions of Mesmerism. He had with him a young man, whose name I do not recollect, but who was a stranger in the place. This man was an excellent clairvoyant.

"One day, while in clairvoyance, Mr. S., the husband of the lady who lost the watch, was placed in communication with him. He inquired of the clairvoyant, who, for the sake of convenience, I will call A. in relation to the disappearance of the watch. For a long time, Mr. A. refused to answer the interrogatories put to him, touching this delicate subject; but at length consented to undertake a full disclosure. His answers were sufficiently definite and descriptive to fasten suspicion upon C. C., a young man who resided in the place, and who had been in the employ of Mr. S., and who had long been a familiar visitor at his house. He stated *definitely* that the watch was now (then) in the hands of a young man in the village of Amsterdam, in the State of New York.

"The credulous, of course, believed that C. C. was the guilty man, especially as he was known to have visited Amsterdam late in the winter. This disclosure was made in the presence of but few witnesses or spectators. The next day Mr. A., the clairvoyant, came to Mr. S., apparently under great excitement, and pointed through the window of Mr. S.'s office, to a young man in the street, and declared *him* to be the young man whom he saw in clairvoyance the day before, and took the watch. The man was C. C., who was a perfect stranger to A. Even the credulity of Mr. S. was now disturbed. He could not, he *would not*, believe the clairvoyant. C. C. had always maintained an unsullied reputation, and Mr. S. had been long and intimately acquainted with him. He was a young man much beloved and respected.

"This young man, C. C., early in the month of August last, was taken sick with a violent fever. After it had raged for a few days with such obstinacy as to preclude the possibility of recovery, he was told by his faithful physician, that his case was hopeless— that he must die! It was an unwelcome message, but he must now be honest, for the scenes of judgment were at hand.

"Two days before his eyes were closed in death, he sent for the Rev. Mr. C., an Episcopal clergyman, with whom he had long been familiarly acquainted. To him he made a free, full, and humble confession of the whole transaction. He disclosed the secret known to none but his God! It was precisely as the clairvoyant had stated it. He took the watch East with him, and sold it to a brother in the village of Amsterdam, as had been stated. He exonerated everybody else from any participation or privity in the affair, and confessed that upon his head alone rested the guilt."

The above *is not, strictly speaking, seeing the past*, as some have been led to suppose. It is nothing more than ordinary clairvoyance and *mind reading*. I have seen many similar instances where clairvoyants have obtained correct information from a third person, *whom they never saw*.

This clairvoyant did nothing more. He received the image of C. C. from the mind of *some one*, and then, by finding him and reading his mind, obtained the information which he eventually divulged. It is a question in my mind, from what I have long since noticed, whether the past may not also be gathered from surrounding objects, even at a distance, by clairvoyants who direct their minds to, or visit them in thought, by a translation of their faculties. I know that they have and can get correct impressions, even by the touch, from inanimate objects that have been present, or in the possession of others.

CHAPTER XIX.

OF THE SENSE OF HEARING.

WHEN the sense of hearing is in this state, the subjects usually do not hear or listen to what is passing around them, unless directed to it by the person into whose care they have placed themselves, or there is an express desire on their own part to do so, and then they hear without any other communication.

When they are desirous of listening, they can translate this faculty to any distance, and hear what there transpires as distinctly as if the thing to be heard were in the same room. This may at this time appear a sweeping assertion, but I can assure the reader that it is nevertheless a truth which the world sooner or later *will have to believe.*

My investigations with this sense have been conducted with the utmost care, and I am persuaded that, if like experiments to the following be repeated by others, the results will be as satisfactory to them as mine have been to me. I select the following from a number of a like nature:

I.—EXPERIMENT.

Subject, a young lady.

Three persons were requested to retire into a distant part of the yard, and to speak of something which they should bear in mind.

When asked what they were conversing about, she said that they were speaking about the kitchen and the piazza, and when requested to state the exact words they were speaking at that moment, she replied that Mr. Z—— just now said that "it will do very well," alluding to the manner of trimming the trees, etc.

They were requested to come in, and were told what she said; and they declared that that was the subject of conversation, and those were the exact words that had been uttered by Mr. Z.

The subject was seated in the front room, and both doors between her and the kitchen, through which they passed, were closed.

II.—EXPERIMENT.

Subject, a lady in the country.

She was requested to state what they were speaking about in the next house, the doors of both being closed, and the distance between them about one hundred yards.

She said they were speaking about a Mr. M——, who lived at a distance. Her statement was ascertained to be correct.

This experiment was performed at the request of a skeptic, on the spur of the moment, without any previous arrangement, and therefore puts the possibility of collusion out of the question.

III.—EXPERIMENT.

Subject, a young lady.

She was requested to state what two other ladies

were speaking about in the next room, who had retired for that purpose. They were directed to go to the farthest end, and to whisper barely loud enough for the one to hear the other.

When they returned, they were astonished to find that not a word had escaped her. The ladies were both skeptical, and before the experiment had been tested by them, they had declared that it was impossible for her to do it.

IV.—EXPERIMENT.

Subject, a lady.

She was requested to listen to some music at a distance of one and half squares. She said that she heard it distinctly, named the tunes that were played and the kind of instruments upon which they were played, etc. This was also performed without any previous arrangement. Her statements were found to be correct.

V.—EXPERIMENT.

Subject, a lady.

She was requested to listen to what a young lady was singing who had been sent into the woods for that purpose by a skeptic, with directions to sing certain pieces merely loud enough to hear herself, and to note which she sang first, etc.

Answer: "She is singing 'My soul is heaven bound, glory hallelujah;'" and after a pause, of perhaps half a minute, she said, "And now she has commenced the 'Promised Land.'" Answer was correct —she sang but two pieces, and those so low that, to use her own expression, "it was impossible for any one to have heard them at a distance of three yards." The woods were one-quarter of a mile distant.

Many individuals who are not clairvoyant often hear, and use this faculty at a distance very well.

I have had many subjects, two in particular, both gentlemen, in whom the sense of seeing was not perfectly in this state at the same time that the hearing was; and who were both enabled to translate this faculty to a distance, and although they could see nothing they could hear what was said or going on distinctly.

They have frequently told what was spoken at the distance of several miles; and when taken to a cocoonery at a distance of four miles, they declared that they could hear the worms feeding as distinctly as if their ear was within an inch of them.

Both gentlemen were skeptical, and entered this state out of curiosity. They have both lately entered the state more perfectly, and are now also most excellent clairvoyants. One entered it perfectly on the tenth, and the other on the twelfth sitting.

With these and many other subjects I have performed many like experiments in hearing, at even much greater distances, and so far as I have been able to ascertain, they have in a great majority of cases been perfectly correct.

If, therefore, when the mind has been properly directed, they can hear the exact words spoken or the tunes played or sung at a distance so far exceeding the powers of this sense in a natural state, how can we limit their abilities?

The same remarks, however, which I made when upon the subject of clairvoyance, etc., also apply to their hearing; and, if their imagination be properly restrained, they can hear things at a distance as well as if they were present and listening with their natural ear. But if they imagine and are careless or indifferent as to the result, their answers cannot be depended upon. The same remarks will apply to the senses of taste, smell, feeling, etc. It is astonishing how sensitive the senses are while in this condition; and subjects are frequently enabled to select from a number of articles, mixed up promiscuously, those which belong to each individual, although they may never have seen the individuals before. This would seem to prove that there was a peculiar impression, or a something, left upon all substances by their surroundings, proximity, or possession, which enables subjects to distinguish between articles owned by different persons. What this something may be, or how subjects discover the difference, they cannot find words to make us comprehend, and simply say, that, by using all their senses, they either see, taste, smell, or feel, etc., their peculiarities, but that it is necessary to fix their minds intently upon them; and when the articles are at a distance, the mind must be cast there, and then the necessary examination made.

CHAPTER XX.

OF THE SENSES OF SMELL AND TASTE.

THE senses of smell and taste, while in this state, like those of seeing and hearing, commonly lie dormant, or inactive, but are at all times under the control of the subjects' will, and they can smell and taste, or not, as they please, independent of any one.

If they *do not* desire to smell, the strongest substances held under the nose are inhaled with impunity; but, if they desire to smell, they can do so with the utmost facility and correctness, and can distinguish the most delicate scents at a distance, notwithstanding the vials, etc., which may contain them are closely corked and sealed.

It does not matter how well they may be secured, or where they are placed, so that the subject is correctly informed of their locality, and the substances to be examined be such as they could name or distinguish in their waking moments. The same is the case with their taste, etc.

I have performed many experiments to prove the powers of these senses, and of which the following are a few of the most interesting:

I.—EXPERIMENT.

Subject, Mrs. H——.

A vial, closely corked, containing some liquid, was placed upon the table, about four yards distant from where she was seated. She was then requested to examine its contents by smelling and tasting it.

After obtaining her consent, she was left undisturbed for perhaps half a minute. I then asked her if she had smelt and tasted it. She said: "Yes; and I know what it is. It smells and tastes like cinnamon."

The vial was examined, and found to contain oil of cinnamon.

II.—EXPERIMENT.

A second vial, closely corked, was placed upon the table, and the same requests made.

Her answer was, "It smells like lemon."

Upon examination, the vial was found to contain a few drops of the oil of lemon.

III.—EXPERIMENT.

A third vial, secured and placed, etc., as the above. "This," said she, "smells very strong. It is hartshorn."

Answer correct. The vial contained spirits of hartshorn.

I have had many other subjects to perform similar experiments with the same success; but it is very seldom that any of them can be induced to perform more than three or four of the same nature at the same sitting. It is therefore best always to vary them, to meet their approbation.

As there is so much sameness in these experiments, I shall give but a few more, which were performed on a different occasion by the same subject:

IV.—EXPERIMENT.

A vial, closely corked, containing a colorless liquid, was placed upon the table. The usual requests were made, and her answer then was, "It smells like camphor."

It was examined, and found to contain spirits of camphor.

V.—EXPERIMENT.

A second vial was placed upon the table, etc. This she said she was well acquainted with; but I had considerable difficulty before I could get her to name it. She, however, eventually said that it was "essence of peppermint."

Her answer was correct.

VI.—EXPERIMENT.

A third vial was placed upon the table, secured like the rest. This she examined for a considerable time, and at length declared that it had "neither taste nor smell."

The vial contained pure water.

VII.—EXPERIMENT.

One of the vials was selected by one of the audience and taken to a neighbor's house, about one hundred yards distant. When I was told where they had placed it, I requested her to cast her mind there, and to smell and taste it at that distance.

She stated that the vial contained essence of peppermint, and that she smelt and tasted it distinctly.

Her answer was again correct.

She now requested me to cease with the experiments for that evening; and, as she had gone through many others of a different nature, and felt disposed to rest, I was obliged to comply.

Had I not done so, and still persisted, the consequence would have been that she would have become careless, listless, and indifferent, and her answers would have been evasive, inadvertent, and unsatisfactory.

This is an unfortunate condition for the successful prosecution of experiments; and, although they always very much regret their not complying when they awake, it seems impossible for some subjects to overcome this feeling or disposition while asleep. This feeling is not natural to all, and is by no means a necessary condition of the state, but is sometimes assumed almost contrary to their sense of courtesy or propriety. I have, however, commonly found them candid; but they sometimes, even when most opposed to performing experiments, show a seeming willingness to do them, yet, if you ask them candidly whether they have looked, smelt, or tasted, etc., as the case may be, they will say no; and if asked the reason, they will say that they did not feel disposed; that they wanted to sleep, or were thinking about something else. It is therefore utterly *useless* to request or urge them to perform any experiment, unless they are perfectly willing or feel disposed to do so themselves.

I will here, still further to illustrate this peculiar condition, give a few experiments in detail, which were but partially successful.

VIII.—EXPERIMENT.

Subject, Mrs. ——.

A wine-glass containing some liquid was placed upon the table, and the usual requests were made.

She was very unwilling either to taste or smell it. Said there was no use in doing so; and that she felt so well and comfortable, that she did not wish to be disturbed. I then explained the reason why I wished her to examine it, and endeavored, by argument, persuasion, and every other means in my power, to obtain her consent; but all that I could do, in the course of half an hour, only wrung from her a partial consent, the nature of which will be better understood by giving her own words; as, "I will think of it; perhaps I may; or, I guess I must," etc. She, however, finally said that she had examined it; but I had the same difficulty in persuading her to say what it was. Eventually she said it was some kind of wine, but could not be persuaded or prevailed upon to name it.

The glass was examined, and found to contain a small portion of currant wine.

IX.—EXPERIMENT.

A second wine-glass, containing a colorless liquid, was placed upon the table, and she was again requested to examine it.

The same difficulty was experienced, and it was a long time before she could be brought to say anything positively. She at length said: "That it had no smell, but tasted very sweet."

The glass contained a solution of loaf-sugar in water.

A third experiment was attempted, but it was impossible to overcome that feeling of listless indifference or independence which had taken hold of her.

The experiment was of course unsatisfactory; not because she *could not* render satisfaction, but because she *would not or could not* overcome the indifference which possessed her.

When, therefore, there is a disposition on their part not to perform an experiment, it is better to drop it at once, as they then frequently say anything to get rid of you. *But I have never yet known them to fail in any experiment when the desire to perform it originated with themselves.*

I shall conclude the experiments in smell and taste by giving two more at a distance:

X.—EXPERIMENT.

Mr. L——, a gentleman who was skeptical upon this subject, requested one of my subjects (a gentleman) to state what was contained in a bottle which he had placed upon a shelf in a certain room in his house, about one mile and a quarter distant.

After the subject consented, I requested him to cast his mind to the place, and to state what was in the bottle referred to. He at once stated that it was "about *half full* of gin."

His answer was correct, "Although," as Mr. L—— afterwards stated, "the contents of the bottle were known to none but himself;" and he had placed it there just before he came over expressly to test his powers.

XI.—EXPERIMENT.

Subject, Mrs. H——.

Having tried quite a number of experiments at short distances, I was anxious to try this lady's powers, which are extraordinary, at a greater distance; consequently I obtained three vials, as nearly alike as possible. I filled the first with spirits of camphor, the second with essence of peppermint, and the third with pure water. All were white and colorless. The vials were then corked, securely sealed, and thoroughly mixed, so that it was impossible to tell the one from the other. In this condition they were given to my wife, with instructions that after I left home she (my wife) was to place the vials promiscuously upon a certain shelf in my office, four or five feet apart, and to leave them in the same position until I returned home the next day. The subject, Mrs. H——, was being treated for a nervous affection which rendered her both blind and lame, but was at this time almost entirely restored through Somnambulism.

My visits to her at this time were made every third day, and as I usually remained all night on these occasions, we had plenty of time for experiments during the evening. I arrived there early, and supper being over, as usual she entered the condition, and after some experiments in clairvoyance which were very satisfactory, I directed her mind to the vials which I had requested my wife to place upon the shelf agreed upon. She stated at once that she saw them, and described their position. I then directed her to cast her mind into the first vial, which stood to the left as she faced the shelf, and then to taste and smell what it contained. After she had done so, she stated that the first bottle to the left "tastes and smells like camphor." I then remarked that I wanted her to be certain in regard to the contents of the vials, as the experiment was an important one, and would settle a great question in my mind. Upon which she again stated that the first vial to the left contained spirits of camphor, the second or middle one, on the right of the first, she examined for some time, and then stated that she saw there was something in it, but that it had no taste or smell. The third, without any hesitation, she declared contained essence of peppermint.

Upon my return home the next morning, to my great surprise I found that her answers were correct. Viz.: that the first vial to the left contained the spirits of camphor, the second or middle one, the water, and the third to the right the essence of peppermint.

The distance between the subject and the vials was about seven miles, and as no one knew how the vials were placed in regard to their contents, or whether they had been placed there or not, the case is a strong test of their ability to taste and smell, etc., at a distance, as could well be desired.

CHAPTER XXI.

OF THE SENSE OF FEELING.

THE following remarks are from a letter to the editor of the Philadelphia "*Spirit of the Times*," dated December 18th, 1843, and published in that paper on the 23d of the same month.

"The sense of feeling when in this state presents many interesting peculiarities, and its study has been rendered particularly important on account of the insensibility which exists, and the advantage which may be taken of this phenomenon in performing surgical operations.

"The possibility of performing operations without inflicting pain has been doubted by many, and the insensibility which exists entirely denied by others. This has arisen from the many apparent contradictions which have been exhibited by different subjects, or the same subjects at different times, or from improper management or a want of knowledge on the part of the operator.

"My attention has been turned particularly to the study of the phenomena of this sense, with the view of obtaining a correct knowledge of its peculiarities, and if possible of finding out the best manner, under all circumstances of alleviating or preventing human suffering.

"I have instituted many experiments to ascertain the facts, and present the following as the most important and interesting:

"When this sense is in this state, and you attempt to inflict pain by pinching or otherwise, they may feel it or not. Sometimes they do, and sometimes they do not. This apparent inconsistency I find is owing to their own will, and they can feel or not, just as they please.

"If you pinch them, and they exhibit no signs of pain, by simply requesting them to do so, or by drawing their attention to it, they can feel as well as when awake, although you may will them to do the contrary as much as you please.

"As, therefore, the subject has perfect control over this sense also, and may exercise it during an operation so as to feel pain, it is necessary to guard against their doing so as much as possible. I have performed many operations lately without inflicting the least pain, or their having the least knowledge of having been operated upon until the fact was mentioned to them.

"The method I pursue, if the subject be clairvoyant, is simply to take or send them in thought to some distant place, or to amuse them by conversing, or otherwise, until the operation is completed. If this be done properly, the operation may be performed not only without inflicting pain, *but they will be unconscious of having gone through it at all.*"

The powers of feeling in distinguishing articles whether by actual contact or at a distance, are as remarkable as those of the other senses, and I have frequently had subjects restore several articles, given them at once, to their right owners, without ever having seen the articles or the owners of them before. This, as I have before stated, would seem to prove that the articles themselves retained a peculiar impression, or a something that enabled the subject to recognize and restore them to their right owners.

They have also told the quality, size, shape, roughness, or smoothness, etc., of articles placed at a distance, or the temperature of solids, liquids, or of the atmosphere in different rooms or places, independent of any previous knowledge on our part, to the perfect satisfaction of those who at different times were engaged in the experiments.

When only a portion of the body is thrown into this state by the subject, say a finger, a hand, or an arm, etc., they still have *the power to feel or not, as they please, in these parts;* but it at first will be more difficult for them to do so than when the mind is also in this state.

The power of throwing any portion of the body into this state, independent of the rest, may be acquired by any person who will practice it under proper instructions; but it will be much more difficult for those to acquire it who have not been wholly in this state, than for those who have; but when they have once succeeded with one part, the rest becomes more easy.

The ability to do this is extremely useful in cases of injury, when the subject, at will, by doing this, could relieve himself from the pain which he otherwise would be obliged to suffer, until a physician or surgeon could be obtained, and the limb or part set and dressed, etc., according to the nature of the injury sustained.

After an operation, or where an injury has been sustained, I always request the patient to wake up, *with the exception of the affected part,* so that no pain may be experienced during the time necessary for its complete restoration.

It is remarkable that, when a tooth has been extracted while in this state, if they have been properly managed by the operator, when they awake they do not miss it; or, in other words, feel the vacancy which has been created by its extraction any more than they would if it had been out for years, and they had become used to its loss; the tongue, as is usually the case, is not thrust into the cavity, and the unpleasant feelings created by its loss is not experienced.

I have lately heard some contend that it is useless to enter this state until an occasion requires that they should. This idea is erroneous and unfounded, because many persons cannot enter it easily at the first trial, particularly when they are laboring under pain, and therefore ought to practice it until they can do so with facility.

Many object to this also, because they fear, that if it be entered too frequently, it may become a habit, and they might fall into it when they did not wish.

This is also a mistaken idea, and all the cases upon record have happened because they have been improperly managed, and did not understand the true nature of the sleep. I have never yet seen anything of the kind, although I have had hundreds under my care who have entered it in less than half a minute, and some of them as often as six or eight times a day, or a hundred times within a year, without any bad effects, or any danger of its becoming a habit which they could not check at pleasure.

To give the reader an idea of the advantage of being able to enter this state quickly, or at pleasure, I will relate the following case of injury, which was entirely relieved in less than two minutes by entering the state at will.

I.—CASE.

Miss ——, in attempting to place a smoothing-iron upon a high mantel-piece, stepped upon a chair, and in reaching up the chair tilted, and she fell across its sharp back, upon her right side, with her whole weight, injuring several of her ribs, etc. Somnambulism was proposed, but objected to, because she did not believe she could enter the state while suffering so much pain. She was then bled, purged, and other remedies applied and used without any relief whatever.

Upon the third day after the accident, I again proposed Somnambulism, but she objected as before. I, however, now insisted upon her making the trial, and, as she had been in the state frequently before, I did not apprehend any difficulty. She was suffering extremely at the time, yet, notwithstanding, she entered the state in about one minute; and, when interrogated respecting the pain in her side, she declared that she did not feel it at all, and kept pressing her side with impunity. She remained in the state about fifteen minutes, when, after being directed to leave that part in the state, she awoke entirely free from pain, and immediately went about her usual occupations.

Two days after I was again sent for, as she had a return of the pain, occasioned by imprudence in over-exertion.

I found her laboring under the same symptoms, and suffering full as much as she did before. She entered the state again, with the same results, and awoke, as before, free from pain, and has never felt anything of the kind, although it is now many years since the accident occurred.

ANOTHER CASE,

Which was related to me by the gentleman himself, is as follows:

Mr. H——, whom I had previously taught to throw any part of the body into this state at will, having had his forefinger mashed between two railroad cars, threw it, although suffering very much at the time, into this state very readily, and declared to me that, from the very moment that he had done so until it was entirely healed, he had not experienced the least pain, although, at the time, he was obliged to press it into shape, etc., until the necessary bandages, etc., were applied.

ANOTHER CASE.

Subject, Miss ———, of Philadelphia.

During the time that I lectured upon Artificial Somnambulism in Philadelphia, I taught this young lady to throw any part of her body into the somnambulic or insensible condition at pleasure, and in about six weeks after I returned home, I received a letter from the father, stating that he would not take a thousand dollars for what I had taught his daughter, as she had lately met with an accident from boiling water, which so severely scalded her leg and foot that the skin adhered to the stocking when it was taken off. Yet, notwithstanding the severity of the scald, she threw the leg and foot into the insensible condition in an instant, and kept it in that state during the time necessary for its restoration, which was not long in being accomplished, as the scalded parts seemed to *dry* up, without any inflammation, pain, or suffering of any kind.

This is a remarkable case, and shows the use of being able to enter this condition at any time that it may become necessary, and as no possible injury can result, or habit arise, from the power of exercising it at pleasure, those who do not avail themselves of its blessings do not only "stand in their own light," but are slaves to prejudice, superstition, ignorance, or bigotry, and unnecessary suffering will exist, until a higher plane is assumed and its blessings realized.

CHAPTER XXII.

OF THE SENSE OF MOTION.

THIS power, like the sense of feeling, etc., I believe to be a distinct sense or faculty situated in the brain, and having all the functions or kinds of action with the rest, and, like them, capable of being educated.

I conceive that, like the rest of the faculties, it must perceive, recognize, or observe, and that this power in this faculty, combined with its own function of love, produces an emotion or a love of motion, which is a passion as well as the love of offspring, the love of order, causes, or comparisons, etc., and that its functions of will—with the will of other organs—having control of the muscular system, can cause all kinds of motions to be made. But when the above three functions, viz.: perception, love, and the will, only act together, they will be motions simply irregular and undetermined, which, however, its judgment, if active, can render definite, or the imagination original. If these be combined or associated with the functions of imitation, imitative motions can be made, and if the memory becomes active, one, both, or all may be recalled, and again perceived and again remade, etc., according as the functions of this faculty are combined with those of another.

If the power of motion acts independent of the judgment of that faculty, then the motions will be irregular, as in Chorea, Sancti Viti, Saint Vitus's Dance, irregular startings, or twitchings, etc. We may desire to move, and yet no motion may follow, as in disease where the faculty is injured, as in Paralysis or Palsy.

In a somnambulic state, the existence of a distinct power to recognize, conceive, and make motions is still more evident, and I have frequently had subjects to imitate my motions when they were not otherwise clairvoyant, and the attention of this faculty with that of imitation was directed to what I was doing, as I have already explained in another chapter.

The imitations, however, cannot be produced when these faculties are not clearminded, or their attentions are not watchful. But when excited, I have frequently seen them continue, at intervals, a long time after all endeavors upon my part to excite them, had ceased. Indeed, sometimes upon entering the state, on another occasion, the same motions would be again made, although I was not thinking of them, or endeavoring to excite like motions in them. In this case, the memories of the faculties which, on a former occasion, were acting, are recalling the former impressions, and the motions resulting are the effects of that recalling. When the mind is otherwise engaged these motions cannot be produced, and they only take place when the attention of these faculties are active, and they may then frequently be produced independent of the subject's knowledge.

When these imitative motions are produced independent of the subject's knowledge, they are the result of the activity of the functions of attention and perception, etc., in the organs of motion and imitation independent of the functions of consciousness, etc., in either of these organs, and are perceptions or mind-readings proper belonging to these faculties.

Some subjects are only partially clearminded, or rather, some of the faculties are clearminded, and others not. I have had some subjects to perceive things and not individuals. Some would see individuals and not colors, etc., according as these faculties were perfectly in the state or not.

I conceive, therefore, as I have before stated in other chapters, that each and every faculty belonging to the brain, when in a state of Artificial Somnambulism, possess this power, viz.: perception, in a greater or less degree, and has power to perceive what relates to its or their peculiar powers, whether it be in the mind of another, or in the external world, etc., and will perceive correctly or not, according as their individual attention has been fixed, or their imagination in the respective faculties may have been restrained or not.

The rigidity or apparent inability to move, which is often exhibited in subjects in a state of Artificial Somnambulism only exists, because the subjects believe that it must be so, and that they cannot alter it. This is a gross mistake, instilled by operators who are ignorant of the facts, and very little trouble will soon convince any person that the subject can, at any time, by an act of their own will, relax any muscle, or make any set of muscles rigid contrary to the will of any one.

They can do this, too, whether the head be in this state or not, and it only requires an exertion on their part to effect either. I therefore contend that all reverse passes, or passes of any other kind, are worse than useless, because the subjects themselves can cause or shake it off in all cases whenever they choose, if the endeavor to do either be really made, no matter whether the so-called operator wills to the contrary or not.

OF THEIR PHYSICAL STRENGTH.

The physical strength of persons while in this state, compared with that when awake, can be much increased by them at will. I could relate many cases in which this has been successfully demonstrated, but it will be sufficient to state that I have seen some hold out at arm's length weights which, when awake, they could not possibly so extend.

Indeed, I have seen some delicate young ladies lift, with apparent ease, weights which the strongest gentleman in the room had considerable difficulty in raising to the same level.

But operators should be careful not to urge their lifting too far, for, although they may not feel the effects when asleep, they might do so when awake, particularly if their minds have not been directed, while in this condition, so as to prevent it. This should always be done before they awake.

CHAPTER XXII.

OF THE INFLUENCE OF ARTIFICIAL SOMNAMBULISM ON THE SYSTEM.

VARIOUS opinions have been entertained upon this subject at different times. Some in whom superstition, prejudice, and ignorance of the phenomena prevailed, were opposed to and spread all kinds of evil reports against it, from witchcraft down to diseases caused by it which could never be cured.

Others, again, who were fond of the wonderful, exalted it to the skies, as being the long sought for desideratum of universal health—the elixir of life and remover of all diseases.

Reports of the question on both sides were exaggerated, and between the two extremes, it was at one time impossible to decide which was true and which was false.

I shall give my experience of its influence, first, upon the healthy subject, and then upon those who are diseased.

I.—OF ITS INFLUENCE UPON A HEALTHY SUBJECT.

If a person in perfect health enters this state, and while in it be properly taken care of and taught how he shall relieve himself, it cannot do him any injury whatever, and he will feel as well when he awakes as before he entered it. He will, if anything, feel better and be more refreshed, buoyant, and active, unless a positive resolution upon his part, while in the state, should produce an opposite result.

If a person should feel weary before he enters it, he will have lost the whole of it when he awakes if his mind has been properly directed, and will be as fit to undergo the same exertions as before he became weary, although he may have only been in it a few minutes. I have frequently, after a hard day's work, seen them entirely relieved in a few minutes from the stiffness and soreness which existed; but it requires that their minds should be properly directed while in the state, or the soreness, etc., will remain when they awake. This will, however, be fully explained when I come to the treatment of diseases.

I have been frequently asked the question, "What good can result from entering this state when in perfect health?"

I answer that, if properly managed while in it, no evil can possibly result, and that it is impossible to say or know how soon the entering this state may become desirable.

Every man is liable to accident, and may become injured or ill, and if he has once been in this state he can enter it with more facility even while he may be suffering pain, and I give this as a reason why all who desire to enter it should do so under the care of proper persons, so that, if necessity requires it, the subject can accomplish it whenever it becomes desirable.

The oftener a person has entered it, the more readily he can accomplish it, and this shows the necessity of practising it until the power to enter it sufficiently soon is completely acquired.

It cannot, however, be denied that injury has sometimes been sustained by some *from improper management*, as the following case, related by J. G. Foreman, while in Lexington, Kentucky, and published in the "*Magnet*," fully demonstrates.

I extract the following from his letter to the editor:

"The object for which I commenced this communication was to relate an accident that occurred with the lad already alluded to, of quite an alarming character, and one that will serve as a caution to persons unacquainted with the nature of the mysterious influence.

"After I left Danville, the lad was magnetized by any one that felt the inclination or curiosity, notwithstanding the warning I gave in my public lectures of the danger of meddling with it without a knowledge of its principles, and of the human system in general. The consequence was that in a short time he was very much injured.

"Persons were allowed to magnetize him on various occasions; and many of them, in exciting the different parts of the brain, handled him very roughly. His mind became considerably affected, and disturbed him in his sleep; and to conclude the amount of injury done him, he finally became DEAF AND DUMB!!

"Several days after this occurrence, I happened to be in Danville again. I saw the lad, and he could neither hear nor speak. He used a slate, and communicated with me in writing. He seemed very much grieved about his affliction, and had already learned the deaf and dumb alphabet, and was beginning to learn signs; he had not lost the memory of words, but his organs of hearing and speech had become paralyzed. I persuaded him to sit down and let me magnetize him properly, and I told him that it would probably cure him. He consented, and in a few minutes he was fast asleep.

"He then, whilst in this condition, gave an account of the cause of his deafness, stating that a physician of Lancaster, by the name of Dr. H——, had enticed him from home—while his brother, Dr. Van Camp, was in Louisville—by false representations to the rest of the family, notwithstanding his brother had expressly forbidden that he should leave home, or be magnetized in his absence; that Dr. H—— magnetized him on several occasions for the amusement of his friends; and in experimenting in phreno-magnetism, had injured his brain by the rough manner with which he handled his head. He also attributed the injury, in some measure, to a similar treatment from others, who had been in the habit of experimenting upon his brain.

"This statement was confirmed by his brother, Dr. Van Camp, and without learning anything more of importance from him, I waked the lad up. As he opened his eyes he was perfectly astonished to see me in the room, asked me when I came from Danville, and talked with me freely as though nothing had happened. We soon discovered, from his conversation, that he was perfectly unconscious of the time he had been in the deaf and dumb state, and upon asking him what day it was, he named the very day on which he fell into this remarkable condition. He had no recollection of having been deaf and dumb, and was astonished at our inquiries."

Many other instances are related by various authors of injury from improper treatment on the part of inexperienced operators, but they never take place where the proper treatment is pursued, particularly if the subjects be acquainted with the true nature of the state into which they are about to enter. They then know their powers, and never submit to any improper treatment from any person.

II.—OF THE INFLUENCE OF ARTIFICIAL SOMNAMBULISM UPON DISEASED SUBJECTS.

If a patient enters this state for the relief of disease, and while in it no allusion to his disease be made, or he does not think of it, or places his mind upon it of his own accord in a proper manner, no relief will be experienced when he awakes.

This fact, so far as I know, has never been mentioned or observed, and is the reason why some persons have entered it without experiencing any relief.

It is therefore *highly necessary*, when relief of any

kind is desired by a subject, that *his mind should be placed upon the disease, and before he awakes* he should *resolve to forget it, or that it shall cease to trouble him when he awakes.*

Early in my operations I observed the power of subjects, while in this condition, to remember or forget what they pleased, or of correcting habits, etc., which were unpleasant, and soon after applied it to the relief of disease; and I have always since found that the firmer the resolution made in this state is, that the disease or habit shall cease, the sooner and more permanent will be the relief experienced when they awake.

Heretofore, most operators have depended upon the sleep itself for the relief of the disease, but I was frequently disappointed in this, and looked for information in vain upon this subject, until I discovered the above method, and have since had but little difficulty when I could get them to fix their minds properly. *It requires much less time and fewer sittings* to effect an object, and I therefore give this as the best method with which I am at present acquainted, and believe that it is the only one in which the desired result can be obtained.

Most subjects have power to create pain in any part of their body while in this condition, and will feel the same when they awake, if they believe or resolve that they will have it before they throw themselves out of the state. It is therefore necessary, to prevent any unpleasant feelings when they awake, to make them resolve to feel well when they have thrown themselves out.

Many imagine they feel badly, or have pains, etc., while in this state, which amounts to the same as if they had made a resolution to have them; and they will suffer as long as this conceit lasts, or until their minds are drawn from it. All unpleasant feelings, however, will subside as soon as the mind is withdrawn or directed to something else, and this the instructor should always be careful to do as soon as they complain. I have taught many who have practiced the art to relieve themselves of pain or disease, even when otherwise perfectly awake. This, however, is not easily accomplished when they have never been in the state.

CHAPTER XXIV.

ARTIFICIAL SOMNAMBULISM CONSIDERED AS A THERAPEUTIC AGENT.

AFTER what has been said under the head of its influence upon diseased persons, it will only be necessary here to add, that the influence of the mind has often been too much overlooked in the cure of diseases in a natural state.

It is well known to every practitioner of medicine, that bread pills, given as a purgative, have acted in that way; pure water as an emetic; and that salivation has been produced by gum pills. We have cases upon record where the hair has turned gray in a few hours through fear; and that even death has been produced by blindfolding, and making a criminal, condemned to die, believe that he was bleeding, although not a drop of blood flowed, etc.

I have cured many affections by simply acting upon the mind of the patient while in a natural state; and among the number were several subject to convulsions, and one particularly who believed himself "*bewitched*," and who had not, according to his own account, slept for six weeks, by simply suspending a leather bag about his neck containing some strong-smelling drugs, and making him *believe* that the cure was infallible. He believed that it would cure him, and the cure was effected, not by any virtue in the drugs contained in the bag, but by his own belief.

I could relate many cases, were it necessary, to prove the influence of the mind in absolute disease, but sufficient has already been said, and I shall now proceed to notice the method employed in Somnambulism by Dr. A. Teste, in his late work upon "Animal Magnetism;" and in order that the reader may at once understand his views upon this subject, I will extract the following:

"Is magnetism alone sufficient for the cure of all diseases? No; and the best proof that can be given of it is, that somnambulists almost always prescribe something more than magnetism. It is, then, beyond all doubt that Mesmer and D'Eslon were deceived when, with their magnetic wand, they effaced the word incurable from the list of our infirmities. I wish to believe that a natural enthusiasm misled these two men; but what would they have done if, more favored by chance, they had discovered the secret of our magnetic treatment as it now exists? if, in a word, they had found that their guide, in each of their patients, was the unerring instinct and sublime reason of a somnambulist?"

Can it be, I must ask, that one of the greatest advocates of the science in France upholds such a doctrine, and depends implicitly upon "the unerring instinct and sublime reason of a somnambulist" for information in the cure of diseases? That he will submit to be guided by the judgment of any of his patients while asleep, and treat them when they awake as they themselves have directed, since it is well known that, whatever they conceit, determine upon, or say of themselves while in this state, will happen to them when awake, because their minds have been so directed or improperly allowed to wander instead of directing them to fix them upon their disease at once, and requesting them to resolve that it shall be otherwise? But he goes on to say: "We shall see in the following chapter what this new medicine is of which we make ourselves the apostle."

"I know not," continued the same author, page 227, "how far the hypothesis which I have laid down occasionally, regarding the medical instincts of primitive men, is well founded; but one thing which I hold indisputable is, that these instincts do really exist at the bottom of all human organization, and that the only state in which these instincts reveal themselves at the present day is the state of Somnambulism. Broussais said, some twenty years ago, to a friend of his, 'If magnetism were true medicine would be an absurdity,' a strict proposition, of which the most distinguished of physicians rejected the consequent, only because he did not believe in the antecedent. Now, I say it, and I proclaim it in the face of the world, this consequent which startled the great systematist of Val de Grace, I admit wholly, explicitly, and without reserve; for the two terms of his proposition equally constitute, in my mind, two undeniable truths.

"All then that remains is to solve this question: Are all patients susceptible of falling into Somnambulism, and consequently of treating themselves? Certainly not. But fortunately, as we have already shown, the medical instinct of a great number of somnambulists may be exercised to the advantage of others. To remove every obstacle then, all that is to be done is to bring the one into relation with the sick persons on whom the magnetism shall have failed to act directly. Thus, to wrest forever the practice of medicine from intelligence and to trust it to instinct, such is the vast project which I conceive; for I tell you sincerely, the clairvoyance of an idiot in a state of Somnambulism, would inspire me with more confidence, if I were sick, than the greatest geniuses which grace modern medicine. And I mean that this new practice of the medical art should be universal, and be applied to all cases. The study of Anatomy, of operations will alone remain in our schools for the purpose of making surgeons; but all the acts of the latter again shall be subordinate to the suggestions of the somnambulist. I know well that, in saying this, I am covering myself with ridicule, because we must not outrun our own age. Jean Jaques has said, somewhere, that it was a sort of madness to be wise among fools. Well, be it so! I am satisfied, if it must be so, to pass for a fool; but I never shall have to reproach myself with the cowardice of having seen an important truth without daring to announce it. My resolution is taken on this point, and I shall follow up my task to the very end.

"We are then now going to pass in review two orders of fact:

"1st. The patients themselves directing their treatment during their Somnambulism.

"2d. The somnambulist directing the treatment of other patients.

"The first question must be treated immediately, the second will form the subject of the following chapter.

"Extatics predicting months beforehand, the return of their accessions, and describing with perfect exactness all the symptoms of their disease, were phenomena which must have astonished, to an extraordinary degree, the first observers who witnessed them. But when they recovered from their astonishment, must not reflection on what they had seen have suggested to them strange inferences?

"Could it not be, in fact, that a patient, so well informed on the causes, the nature, the course and issue of his disease might know something regarding the expedients to be employed to cure or relieve him? Certainly, such an idea could scarcely fail to occur to a physician, however infatuated he might be with his profession; and if the patient answered, if he suggested remedies and traced out a plan of regimen, would there be a reason for wavering about conforming to his advice? For my own part, I think that my medical pride would not hesitate about bowing before these prejudices, and that I should be profoundly ridiculous if I attempted to write a prescription for this new oracle of Epidaurus, who foretold, a month ago, a disease the existence of which I would not have suspected one hour before its invasion. What! when his life is at stake, and consequently he can have no idea of deceiving me, when he assures me that he knows the remedy required for his disease? As he knows the causes and nature of his disease should I dare to give him my advice and mix my voice with his? Oh, no! I am silent. I give up my rights; I renounce my poor miserable knowledge, and I bow with admiration before those sublime revelations which must emanate from God himself. I hear his prophetic voice to register, with minute exactness, all the words it utters; then to follow, step by step, the counsels I have received from it. To this alone shall I confine my functions.

"Now, what is the result? That under the influence of his own prescriptions, this future demoniac obtains a rapid cure. Well, now let us conclude he is cured. He is cured by means of which I should not have even thought, by a strange mode of treatment, the idea of which would never have occurred to me. His medicine then is the true one; but then, what would mine have been?"

I was aware that the science of Artificial Somnambulism was not properly understood in France, but

I was not prepared to hear such erroneous doctrines from that source.

I believe Dr. Teste to be honestly sincere in all that he has said, but he has been misled by appearances, and has taken too much for granted. He saw that they predicted their own cases, etc., correctly, and, without looking for a cause, seized upon it as a guide to his future operations. Hence his theory and blindfold application. It is no wonder that the science should suffer and meet with ridicule at every turn, when the leaders of the science publish such palpably erroneous and undigested doctrines.

The "medical instinct," upon which so much stress is laid by this author, exists only in the "mind's eye" or imagination of the doctor and his subjects. The predictions, or rather what has been determined upon by subjects months before while in this state, I am well aware, come to pass—but *why do they come to pass? Simply because they have determined that they shall, and believe that they must and will.* But it does not follow, that because these beliefs or determinations do come to pass, that they are true knowledge and a "prophetic voice," or "revelations from God himself," or that the cure might not have been effected in one tenth the time if their minds had been properly directed, instead of trusting to the "clairvoyance of an idiot," or the caprice of any other subject.

I deny, most positively, that they have any more knowledge of medicines in this state than they have when awake, and again insist *that if they never had been suffered to predict, what did follow their predictions never would have taken place.*

If the subject has been properly informed respecting the true nature of the phenomena of this sleep, no such predictions will ever be thought of; and *I contend that much injury has been sustained and much valuable time lost by suffering subjects to believe that they can predict or prescribe in this way.* These predictions should never be permitted, and all who instil such doctrines and permit such notions to be practised, I consider *the cause* of all the ills that may follow.

To give the reader an idea of the manner in which these predictions and prescriptions are brought about, I will quote a case from the same work. It is as follows:

"It is precisely three o'clock. Josephine submits with confidence, and is asleep at ten minutes past three.

"'Are you asleep, Josephine?'

"'Yes, sir.'

"'Are you sufficiently magnetized?'

"'Yes, sir; but when you pass your hands over my chest you do me the greatest good.'

"I magnetized her for some minutes over the region of the heart, and she says she feels a calm and an agreeable sensation.

"'Do you now think that magnetism will cure you?'

"'Yes, I am certain of it, nor will it require a very long time.'

"'How long will it require?'

"'I do not know yet; I shall be able to tell in a few days.'

"'You do not see clearly?'

"'No, but I shall soon. Wait. I shall see clearly to-morrow.'

"'At what hour?'

"'At three o'clock. No—at a quarter past three.'

"'Will you then be able to tell us what we should do to cure you?'

"'Oh! yes, I shall tell you.'

"'How long must we let you sleep?'

"'Till within a quarter of four.'

"'What is the time now?'

"'Twenty-five minutes after three.'

"I looked at the clock, to which Josephine's back is turned, and she is perfectly right as to the time.

"'How will you be this evening?'

"'I shall be very well.'

"'And during the night?'

"'Very well, indeed.'

"'Will you have an appetite for dinner?'

"'Not much, but I must eat for all that.'

"'What?'

"'Soup and beef.'

"'You told me that you digested soup with considerable difficulty, and that the meat always disagreed with you.'

"'True, but this evening it will do me no harm.'

"'You must then recollect this when you awake.'

"'Yes, sir. Oh! I beg you,' adds she, 'magnetize me a little more over the heart—that does me good.'

"I submitted to Josephine's desire; she thanked me several times in the most grateful terms of acknowledgment.

"Some minutes after, I awakened her just at the time she mentioned.

"She smiles at awaking, as at going to sleep. Her looks at first express dullness, then astonishment, then comfort and gratitude. She rises in a sprightly manner, and cries out with enthusiasm:

"'It is astonishing how much better I feel than I felt some days ago; I feel myself as light as a dancer.'

"I impress on Josephine the obligation imposed on her to take soup and beef for dinner, and after having given me her formal promise to conform in every particular to my directions, she leaves me, and goes down stairs running."

After what I have already written in this work, it will be unnecessary to make any further comment, as I presume my readers will see the absurdity and uselessness of the above proceedings.

I quote the following case of ecstacy and catalepsy to show that valuable lives have been put in great danger, and much time lost, by suffering them to predict and prescribe for themselves, and then enforcing their own prescriptions upon them when awake.

"Case of Madame Comet," from Dr. Teste's work, page 230.

"Nov. 25th, 1839, Madame Comet predicted, in the presence of several members of the Academy, that Dec. 5th she should be seized with a stitch in the side, and that, without any reference to the period of her menses, it would be necessary to bleed her. Accordingly, the day before yesterday she was attacked with a severe pain in her left side. In her last sleep she stated that this pain is seated in the lung, that there will be spitting of blood, and that to-morrow at nine o'clock in the morning, *it will be necessary to bleed her to twenty ounces.*"

As M. Frapart's letter, which corresponds to this phase of the disease of M. Comet, includes a multitude of little details, the recital of which we cannot abridge without altering the truth, we shall transcribe this letter without altering the text.

"To Monsieur Bazile. A Courquataine.
"PARIS, Dec. 16th, 1839.

"MY GOOD FRIEND:—I take up the history of the disease of Madame Comet at the moment this lady is just after losing twenty ounces of blood. It was the 8th of the month. Since then, every evening, Madame Comet has an accession of Somnambulism, which lasts sometimes a quarter of an hour, during which everything occurs just as in that which I have described to you; that is, it presents two very distinct, successive states; one of ecstacy, and the other of catalepsy. In the latter, the patient *appears* to hear nothing, see nothing, feel nothing, comprehend nothing; does not speak, stirs not, scarcely breathes, retains immovable all the positions given her, and, I hardly dare say it, *seems* to have lost a portion of the weight of her limbs. In the former there are other strange phenomena. The patient finds herself, I mean *has the air* of finding herself, in communication with a being whom nobody sees, no one hears, no one touches; and to whom, however, if a serious man may be allowed to narrate such impressions, one would be almost tempted to believe, she speaks and answers. The first of these facts is extraordinary, the second is astounding! It is in this state of ecstasy that Madame Comet speaks of her disease; says *where* it is, *how* it will go on, *when* it will terminate, orders the treatment suitable for the fluxion of the chest under which she labors, does not forget the regimen, prescribes the dose of opium which ought to be given to her, predicts the hour of the duration of her accession of the following day; determines, in fine, the day when she will no longer have any accessions.

"At each sitting it is just the same thing, with some variations, depending, no doubt, on the course of the disease, and which I shall notice as I go on. It was during the crisis of the 8th, Madame Comet states, that the twenty ounces of blood taken from her in the morning were inconsiderable, and that it would be necessary to take a pound from her again the day after to-morrow. We weigh the blood drawn, and we actually ascertain that the measure prescribed was not obtained. If it is for this that we must re-commence the operation, it is rather disagreeable, and even a little alarming; for the disease is so old, and the patient so weak, that further depletion may prove fatal. Besides, supposing the prescription infallible, how are we to keep clear of some slip or oversight that may occur in the execution of it? This seems to me very difficult. In the practice of our profession, it is only by an exception that even the most skillful can obtain their end accurately and correctly. It is deplorable, but so it is. Decidedly, Madame Comet is in a bad way; and, however learned her physician may be, however devoted her attendants may be, I am uneasy regarding the result. I think it will be difficult to reach the harbor in perfect safety. However, as we have no reason to distrust the prescriptions of the physician, they are made up to the very letter. Accordingly, on the 10th, after all the precautions taken beforehand, M. Comet takes from his patient nearly seventeen ounces of blood. This time, at least, we do not err by deficiency. The fact is, that in the course of the day the symptoms of fluxion of the chest diminish, and in the accession of ecstasy in the evening, Madame Comet assures us that all is going on better; that all is well; that everything has succeeded. On the next day the same assertions on her part—the same security on ours. But it is all mere chance in this world. On the 12th the patient declares that another bleeding will be necessary to destroy entirely the pulmonary inflammation; that this bleeding is not to be performed either on the 13th or 14th, but on Sunday, the 15th; that we shall hesitate to perform it for her; and that she cannot yet determine the quantity. Such a prediction puts us to a nonplus. M. Comet is not so formed for passive obedience as to be able to walk with his eyes shut; and as for me, though a little more pliant—at least with facts of this kind, considering that for years back I practised homœopathy—I have got out of the way of butchering my patients—I felt almost disposed to doubt and kick. But, all at once, recalling my long experience, *which taught me that a somnambulist, no matter what he prescribes for himself, never prescribes anything wrong; since he is always safe when his prescriptions are accurately followed*, and my profound ignorance of the secrets of nature, I submit, and endeavor to induce M. Comet to do the same. At length, submission. During the storm, it is better to accept as pilot the first that offers than to take none. It is taking at least a chance of safety.

"On the evening of the 14th, Madame Comet, who, no doubt, up to this did not wish to alarm us, tells us it will be necessary to take from her *full twenty-four* ounces of this precious liquid which sustains life; and that, even if she is weak, the bleeding must not be neglected, for *syncope is necessary;* without this, that matters cannot be concluded, or, rather, they would be concluded very soon.

"M. Comet staggers; his poor patient is so ill; she is so weak, so pale, so bloodless, so sunk, so dying, that in truth one must have a stupid faith, or a deep-rooted conviction to venture to continue a course which appears beset with so many rocks. For my part, however, my resolution is formed; true it is, it is not my own wife I have to butcher in this way; and yet, if it were my wife, I am convinced I would not flinch. No somnambulist was ever a suicide. In the midst of a dark sky have we not a star to direct us, and which will not disappear until we shall have no further occasion for it? But if this star should happen to fail us before the time! O darkness! darkness! that would be dying alone in the catacombs!

"Amid the hopes and fears which disturb our minds, after having taken all our dimensions so as not to go either to the right or to the left of our destination, so as not to remain short of the goal, or to go beyond it, yesterday, at nine o'clock in the morning, M. Comet performs a large bleeding, in which the blood flows quite freely; nearly twenty ounces were taken! and we see no syncope come on. The arm is tied up; but scarcely is the bandage applied when alarming circumstances appear; all present become alarmed at them; they, however, are ultimately appeased; I then leave the patient; twenty minutes after, new symptoms arise; all present are terrified, and commence to weep; they run up to me as if I could do anything. There am I the doctor again in spite of me. But what part am I to take when there is no part to be taken? But, instead of pretending to cry, as every physician who knows his business should do in such circumstances, I endeavor to set a good face on a bad game; I encourage the disconsolate family by saying to them: 'We are not mistaken; the somnambulist is never deceived; let us remain calm. Besides—hope did not yet abandon me—have I not myself been bled eight times in one and the same disease, not to mind the several hundred leeches besides? and I am not dead. Then my principle is not to despair of the game till it is absolutely lost. Madame Comet is not dead!—she will not die!'

"However, the day passed on amid great anguish of mind, as it always did at the hour it should occur. There are cruel efforts to vomit; some hesitation is felt about giving a large dose of opium—there is but one instant for the seasonable administration of this disgusting draught. In short, the accession does not arrive—the star no longer shines—we are out of our latitude. I arm myself with courage, and fly for refuge to my conscience. However, as luck would have it, the accession is only retarded—there it is! 'All has passed off quite well,' said the patient, in her ecstatic sleep; 'the bleeding has not been too much. Give me instantly the dose of opium I was to take. To-morrow the stitch in the side will be diminished; and next Wednesday I shall be quite freed from it. With respect to my accession, their disappearance is for the 28th of this month. I am weak, no doubt, and shall be so for a long time. My convalescence will be painful. It is necessary to begin to give me nourishment in order to recover my strength. The food I shall point out will do me no harm. To-morrow, at half-past eight, my accession will come, and will last fifteen minutes; the same quantity of laudanum will be given to me as to-day. Thank God, it is gone!' Then comes on the cataleptic state, which is soon followed by the waking state. And I too awake, and am much relieved. I feel, for I had nightmare, the life of a woman was lying heavy on my chest.

"Fortunately, in great crisis, the depth of the abyss is not measured until it is quite cleared.

"Adieu, etc.,   FRAPART, D. M. P."

With respect to the above case of ecstacy, it is very evident that her predictions made the bleedings and enormous doses of opium necessary; but, if she had known the true nature of her state and her absolute powers therein, it is also very evident that she never would have predicted.

If, instead of predicting and prescribing for herself, she had resolved to throw off her disease, she could have done so at once, without the bleedings or the opium, etc.

Her recovery must have been very slow, for in April, she was still in a state of "great debility;" and, to say the least of it, her life was certainly trifled with.

As is usual in such cases, her predictions were all verified; and as their predicting and method of fulfilling them is the same in cases of Artificial Somnambulism, *which is truly the same condition*, I consider it unnecessary to quote a case of the latter description to illustrate it further. It will be sufficient to say, that in the case of Madame Teste, which occupies about twenty pages of the same work from which the above is taken, and who, feeling an indescribable illness, entered this state for the purpose of finding out something respecting her disease, and upon whose predictions the same dependence was placed by Dr. Teste, her husband, and who fulfilled her prescriptions with the same fidelity, regardless of anything else, at the imminent risk of her life; and, although she eventually recovered, I must still insist that, if she had not been permitted to do the one, the other would not have been necessary.

Many similar cases are related in this work, and, as it has been translated into the English language and introduced among us, I deem it my duty to warn others from falling into the same mistakes.

Some, I am sorry to see, have already fallen into the error, and have published cases in this country. I hope, however, what has been said in this work will soon dissipate the evil, especially when it becomes generally known that subjects can, when properly instructed, throw any part of their body into an insensible condition, and keep it in that state as long as they please, or until restoration has taken place.

CHAPTER XXV.

OF THE KINDS OF DISEASE CURED WHILE IN THIS STATE.

WITH respect to the kinds of disease which are cured or relieved by Artificial Somnambulism, or by the method I have proposed while the subject is in that state, I have but again to say, that whenever irritation or pain is a cause or symptom of disease, that that irritation or pain can be relieved by entering the somnambulic state, no matter what the cause may be; and that both must cease so long as the body or the diseased part remains in the insensible condition; and, if the mind of the patient be properly directed while in the state, the relief will be permanent when they awake; or they can positively be taught to feel or not, as they please, when perfectly awake.

By this statement, the reader will observe that I am not prepared to say that it will cure everything, but that, like all other remedies presented, *it* can never retrieve an absolute loss, nor restore a function virtually destroyed; yet I have known it to relieve many cases of disease after repeated courses of medicine and the laying on of hands, etc., had entirely failed. Pain can always be relieved. As a forlorn hope, therefore, it deserves the attention of all who are interested in the cure or alleviation of human suffering.

The diseases relieved by the method which I have proposed have been both of an acute and chronic nature; such as, Inflammations, Inflammatory Rheumatism, Erysipelas, Scarlatina, Nervous affections, Chorea, Amaurosis, Hysteria, Melancholy, Nervous or Sick-Headache, Dyspepsia, Fevers, Fever and Ague, *Labor-pains*, *After-pains*, and a variety of other local pains from injuries and other affections.

I will now give a few of the most important and interesting cures, in detail, which have been effected in persons while in this state under my care.

I.—CASE—*Chorea, or St. Vitus's Dance.*

Miss H——, aged 10 years, had been under medical treatment for six weeks before I was called in, and had been gradually growing worse.

I found her seated upon a large rocking-chair, propped with pillows and bolsters, unable to speak a word, to stand, or to use any of her limbs at will. She had lost control of her muscular system entirely, and when her eyes turned one way it was impossible for her to bring them back, or to fix them upon any object, even for an instant, and they kept rolling from one part of the socket to another, sometimes becoming fixed in one part and sometimes in another, etc. Her arms, in like manner, were twitching and starting involuntarily, and it was impossible for her to hold anything in either of her hands, but more particularly the left.

The involuntary contractions of the muscles of the face and forehead were frequent, and produced many singular and grotesque expressions, which it was out of her power to prevent.

I prescribed medicine for about two weeks, without any evident advantage, and then resolved upon trying Artificial Somnambulism, the nature of which was explained to her, and her consent obtained to make the trial.

She succeeded in entering it but very partially at first, but after repeated trials she entered it more deeply, and then began to improve rapidly after every sitting; and at the end of six weeks could speak, stand, and walk with considerable ease. She had but one sitting each week.

Several weeks after, I was again called to see the same patient, who had a partial return of her former disease. She had been from home some weeks, and had taken a violent cold, which was followed by considerable twitching, particularly in the left arm and side of the body, etc.; and as some of the family had attributed her former cure to the medicine which I had first employed, I resolved, with their consent, to give the same medicine a fair trial, and accordingly prescribed as I had done in the first instance, and continued for the space of six weeks, but without the slightest improvement on her part.

Somnambulism was again resorted to, with the

same effects as before. An immediate improvement followed, and after the fourth sitting was discharged cured; and, although many years have since elapsed, there has been no return of her disease.

II.—CASE—*Epilepsy.*

Miss E——, aged 19 years, was subject to epileptic fits at intervals of from two to four weeks.

She succeeded in entering the state perfectly upon the second sitting; was a very good clairvoyant, and was perfectly relieved. She had but four sittings, and has never had any return of her disease. Five other cases have been cured in the same manner, one of which was of twenty years' standing.

III.—CASE—*Dyspepsia.*

Miss —— had been laboring under dyspepsia for the last three years, attended with various irregular pains, particularly in the side, stomach, and back.

She entered this state upon the first trial. Was a good clairvoyant, and had several sittings for experiments after she was entirely restored. Her sittings in all numbered twelve or fourteen.

Many other cases have been cured in from one to five or eight sittings, accordingly as they entered it, perfectly or not.

IV.—CASE—*Intermittent Fever.*

Mr. S. P. B—— was subject to ague every third day for six months. He entered this state but imperfectly upon the first sitting without any relief.

On the second sitting he entered it more perfectly, about half an hour before the chill was expected.

He remained in the state about two hours, during which time his mind was diverted by directing him to cast it to distant places, and to see clairvoyantly what was there going on, etc. At the end of that time, no chill or fever appearing, he was requested to awake, after his making a resolution that he would forget his disease.

I met him five months afterward, when he declared that he had felt nothing of the ague since he had been in the sleep. Many others have, from time to time, been relieved in the same way.

The idea that intermittent fever always requires a specific remedy for its cure, seems to be dissipated by these facts, and that, like other diseases, it can also be cured simply by abstracting the mind from the body, or forming a resolution, while in this state, that the difficulty shall be forgotten or cease to annoy. Or again, by keeping the subject in this condition until the time for its accession shall have passed, thus breaking up the habit, and with it, the recurrence of the disease.

V.—CASE—*Fever.*

Miss A. P—— was seized with a high fever, accompanied with violent headache, giddiness, and restlessness in October, 1843, which continued unabated for three days and nights.

I was not sent for until the evening of the third

day, and not being at home, word was left for me to visit her the moment I returned. I returned from the country about one o'clock the next morning, and visited her immediately.

I found her laboring under a high fever, was very restless, and described her head as being "ready to split" with pain. As she was *very much* opposed to taking medicine, and had often, out of curiosity, been in a state of Artificial Somnambulism months before, I proposed that she should enter that state. She at first objected, as she said it was impossible for her to enter it as long as her head ached as much as it did at that time.

I told her she had but one choice besides, and that was a dose of medicine. The thought of medicine decided the question, and, after a third attempt, she threw herself into it in less than a minute.

Upon asking her how she felt she said she was somewhat relieved, but still felt the pain along the side and back part of her head. I directed her to throw her mind upon something else, and not to think of her head; and, as soon as she had done so, she was entirely relieved, and declared that she did not feel a particle of pain.

Five minutes had not elapsed since she was awake, sick, and suffering torments; now, she was well, lively, and, as usual, in health, began to laugh and talk as if nothing had been the matter with her, occasionally joking about the medicine, saying that "this medicine (viz.: Somnambulism) is very easily taken, and I shall, hereafter, prefer it on all occasions."

She remained in the state about half an hour, and after directing her to forget or throw off her disease, I requested her to awake, with the understanding that she should remain well when she did awake. She awoke perfectly relieved, and in fine spirits.

I saw her during the day, and found her as I had left her in the morning—well, sprightly, and ready for her usual vocations. She never had any return of the disease, and the only thing that I regret, in connection with her case, *is* that the whole world did not witness, as I did, *the triumph of mind over positive disease.*

VI.—CASE.

Mrs. S—— had been laboring under a pain in her left side, in the region of the spleen, for the last six years, which was caused by an injury received from falling off a shed roof, upon which she had been engaged in hanging up clothes. She fell off backwards, and fell upon her left side. She suffered extremely at the time, and although she had applied to many physicians, and used a great deal of medicine for several years, she could obtain no relief.

She was, at length, induced to try Somnambulism, and although she had never been clairvoyant, she entered the state perfectly in every other respect, and improved rapidly under the proper instructions.

After entering it several times she was entirely relieved, and has remained perfectly well ever since, now many years ago.

I ascribe her not seeing, while in this state, to her unwillingness to look, and when awake she has frequently acknowledged that she felt so well while in the state, that she did not care about looking, and always disliked to be disturbed, preferring much to be let alone, so that she could enjoy the delightful sensations which thrilled through every nerve.

VII.—CASE—*Inflammatory Rheumatism.*

Mrs. B—— was afflicted with inflammatory rheumatism for two years, during which time all that medical skill could prescribe was used in vain. She gradually became worse, until all her joints became stiff, so much so, that she could not extend her limbs, nor bring her hands to her mouth. The joints of her fingers were much swollen, stiff, and painful.

When I first saw her, besides the condition of her limbs, she suffered intense agony in various parts of the body. She was unable to turn in bed, and was obliged to awaken some one when she desired to change her position during the night. In the daytime she had to be carried from place to place, and fed like a child.

I was called in as a physician, and requested to prescribe medicines for her relief. This I did to the best of my abilities, but all in vain; everything that I could devise seemed to have no beneficial effects, beyond temporary relief; and, as all other remedies failed to produce the desired result, I proposed Somnambulism, which, after the necessary explanations on my part, and a due consideration upon theirs, was eventually agreed to.

After a few sittings, I succeeded in teaching her to throw herself into the state perfectly. A *great improvement* in regard to her pains, etc., was immediately experienced. She was also soon able to turn in bed without assistance, and to use the upper extremities so as to feed herself, and to push herself about when placed upon a chair with wheels.

A gradual improvement also took place in her lower extremities; and when I last saw her, she was able to stand and walk with crutches, the knee-joints gradually yielding and the limbs becoming more straight.

Her pains have long since ceased to afflict her, and her perfect recovery is but a matter of time.

This was a remarkable case, and I cannot find words to convey a correct idea of the crippled, helpless, and suffering condition of this lady when I paid my first visit; and now, to see the change, I can scarcely realize that she is indeed the same person. Her looks proclaim that she is happy in the change, and Somnambulism, the humble agent of the mind, has gained another sterling victory.

VIII.—CASE—*Chronic Rheumatism.*

Mrs. E—— was afflicted with local rheumatic pains in her head, back, and limbs. She entered the state upon the first sitting, and was an excellent clairvoyant.

One sitting relieved her entirely; and I may safely say that all other cases of a similar nature will also be relieved by entering the condition perfectly.

I have never yet seen a case of rheumatism that was not relieved by entering this state; but the permanency of the cure will, of course, depend upon the care and prudence of the subjects themselves, as the cause which produced it orignally may produce it again; but, when a person has been in this condition once, the remedy is always at hand.

IX.—CASE—*Hysteria.*

Mrs. G—— was subject to frequent attacks of hysteria, headache, and other local pains from her youth. She entered this state upon the third trial, and after four other sittings was perfectly restored.

She has had no return since she first entered the state, now seventeen months, and will remain free from attacks as they were originally, until there is cause sufficient to produce them again.

X.—CASE—*Melancholy from Unrequited Love.*

Miss —— had been melancholy and desponding for several years. She was induced to try Somnambulism for her relief.

She entered the state perfectly the first trial, in less than ten minutes; and after she had been in it for some time, I asked her, as is usual in such cases, whether she did not think that it was better for her to forget an attachment which could not be returned?

She said: "*Yes, I believe it would.*" I then asked

her whether she was perfectly satisfied to do so, and to become lively and happy hereafter. She said: "Yes; and I am resolved that it shall be so." With this understanding I requested her to awake.

She awoke and retired with a friend. I have since been frequently informed that she has banished the circumstance from her mind entirely, and has been lively, contented, and happy ever since.

XI.—CASE.

About ten years ago, Mr. J. H—— lost his left eye and was otherwise much injured by the premature explosion of gunpowder, and ever since that time had been affected with rheumatic pains in various parts of his body, and a severe, incessant headache, which rendered life a burden to him. He had consulted many physicians, and had taken medicines from each without the slightest benefit.

As a last resort, through the recommendation of others, he resolved to try Somnambulism.

He entered the state perfectly upon the tenth trial, and has ever since—now about three years—been entirely free from all pains, and has frequently since declared to me that he feels like a new man.

This gentleman is one of the *very best clairvoyants*, and has since frequently entered the state for phrenological purposes, and I am indebted to him for many accurate and interesting experiments.

XII.—CASE.

Miss Z—— was very delicate and small for her age, sickly, nervous, with a sallow complexion and always subject to violent headaches, which were frequently accompanied with flying pains in her ears, teeth, jaws, arms, sides, etc.

She entered this state perfectly upon the third trial, and improved in complexion and general health *very soon*; losing all her pains immediately, and becoming lively and active. She is now, to use the language of her father, "grown to a strong, healthy, buoyant, and active young girl;" and although she has not been in the state for the last two years, her improved state of health remains unimpaired.

She is *one* of the very best *clairvoyants I have seen*, and possesses a fine mind. I am indebted to her for many correct experiments in *seeing, hearing, feeling, taste, smell,* etc.

XIII.—CASE.

I was called upon by Miss ——, residing in the country near the city of Lancaster, to know whether anything could be done for an affection of both arms which resulted from an attack of acute rheumatism which she had been afflicted with about a year before.

Upon an examination I found that she was unable to straighten either arm, and when the attempt was made, it gave her considerable pain.

She stated that she had been unable to use them since her recovery, and was fearful that they would remain so, as she had consulted several physicians and employed many remedies to no purpose.

I immediately proposed Somnambulism as the only

remedy that I thought would be of any use, and stated that if she could succeed in entering the state, that I could relieve her in a very short time—perhaps immediately.

She consented to make the trial, and entered the state in about fifteen minutes.

After satisfying myself that she was perfectly in the state, I requested her to extend her arms, which, with a slight assistance from me, she was enabled to do perfectly after a few trials, and upon asking her whether she experienced any pain when she did so, she replied, "No, not the slightest."

She remained in the state about an hour and a half, during which time, between her clairvoyant excursions and lookings, I frequently requested her to extend her arms. At the end of that time, after impressing upon her mind the necessary resolution, I desired her to awake. She did so at once, and to the surprise of herself and the friends who came with her, she could extend her arms as well as ever without the least pain. I have heard from her often during the last five years, and she is still able to use her arms as well as ever.

XIV.—CASE—*Contraction of the Muscles of the Fingers.*
(Flexor Sublunis Perforatus and Flexor profundus perforans.)

Miss ——, from the country, aged 20. Subject to epileptic convulsions from her youth.

I was consulted respecting the permanent contraction of the muscles of the fingers of the right hand, which had taken place gradually during her convulsions. For the last year the contraction of the flexor muscles had been permanent, rendering the hand, with the exception of the thumb, immovable, and entirely useless.

She entered the state upon the first trial, and while in this condition was requested to extend her fingers. She succeeded in doing so to a considerable extent, and after several other sittings accomplished it entirely. She is now enabled to use the hand as well as ever. Her intellect had also been much enfeebled by her disease, and during her sleep she was instructed to improve it. Her mind was more collected when I last saw her, and the convulsions so far conquered as not to "*break out.*" She intends during the coming spring to sit again, and may then be enabled to throw it off entirely.

The above was written in 1843, and as the subject never came according to promise, I lost sight of her, and therefore cannot report the ultimate facts in her case.

XV.—CASE—*Scarlet Fever.*

Miss K. Z—— was taken down with scarlet fever four days before I was consulted, and when I visited her, I found her laboring under a high fever, attended with considerable pain and soreness of the throat, and a scarlet blush over the whole body.

As she was very much opposed to taking medicine, she asked me whether I thought Somnambulism would have any effect upon her disease. She had often been in the state before, at first out of curiosity, and afterwards for the purpose of having several teeth extracted.

As she seemed to prefer Somnambulism to taking medicine, I told her she might use her own pleasure in regard to it, but that I had not yet had an opportunity of testing its virtues upon cases of a similar nature.

Accordingly she threw herself into the condition in a few seconds, and when asked respecting her throat, she declared that the pain was entirely gone. She remained in the state about an hour. At the end of that time the fever had abated, and when she awoke, she was entirely relieved of all unpleasant symptoms, and had no return of her complaint whatever, after that time.

This young lady is *one* of the *very best* clairvoyants and mind readers I have ever seen.

XVI.—CASE.

On the 20th of January, 1842, seven weeks after Mrs. H—— was confined with her second child, she was taken with a pain in her head and eyes, which was much increased upon stooping.

During the day she was annoyed with a twinkling or a zigzag motion of the light before her eyes, and as evening approached, her sight began to fail, and by the next morning she was completely blind of the right eye.

The left was, at this time, still good, but in about eight days after it also began to fail, and, at the end of two weeks, she could scarcely distinguish day from night. Her eyes remained pretty much in the same state, sometimes rather better, and then again worse, until June, when they became so bad that she could not, at times, make out to know her most intimate friends.

About this time her mouth and stomach, as she said, "appeared to be raw," so much so, that she was obliged to abstain from eating or drinking anything that was the least strong; even very weak coffee occasioned great pain in those parts.

She also frequently complained of a numbness, or a dead feeling which she had experienced in her toe, gradually extended to the same foot and leg, and soon after to the right toe, and gradually growing worse in that also, until the first day of February, when she was unable to move her legs, feet or toes at all. She had also, for some time back, experienced the same deadness coming on in her fingers, but as this feeling seemed to grow worse in her limbs, her eyes became better. She had feeling in her limbs, but never complained of any pain.

Several physicians attended her during the progress of her disease, but as she seemed to be growing worse instead of better she became alarmed, and hearing of the effects of Somnambulism in similar cases, she resolved to give it a trial. I was accordingly consulted soon after, and visited her about the middle of February.

At this time, she was at the house of her father, about seven miles from the city, where she had taken ill while on a visit.

I found her seated upon a large rocking chair, and, after receiving the above history of her case, I examined the symptoms as they existed at the time.

I found that her eyes had improved considerably since her limbs had become effected, and, at that time, she could distinguish persons, but could not see the largest hands upon the clock immediately opposite to her.

She was unable to move the lower extremities in the slightest degree, although she tried a considerable time to effect it at my request, and was obliged, when she desired to move them, to do so with her hands.

They were obliged to carry her both to and from bed, and whenever she attempted to stand, when held up, her legs refused to bear her weight, and, as she expressed it, "broke down," and, if not supported, she would have fallen to the floor. She was extremely nervous and complained of a general weakness, both of body and mind.

After the necessary explanations had been made to her respecting the state she was about to enter, and the directions given how to proceed, the trial was made. After thirty-five minutes had elapsed, as I found there was no likelihood of her entering it perfectly, I requested her to cease until after tea, which was just then announced. During this sitting, however, her fingers and arms twitched considerably, but her legs were not effected. *No relief followed this trial.*

As I was anxious to know whether she could move her toes, I requested her to move her stockings, so that I could ascertain to a certainty whether there was any motion, when she made the exertion to move them.

I could not, however, perceive the slightest motion, although she made many and often-repeated trials to do so. I was now satisfied that if an improvement did take place, we should be able to see it at once.

After tea I again made the necessary explanations, and dwelt particularly upon those instructions which she had not observed in her first sitting. She stated that the noise created by those moving about the house had disturbed her. After enjoining silence upon all, she made a second attempt.

The twitching this time, extended to the lower extremities, and an occasional start in the muscles of the left leg and foot assured me that some effect would follow this sitting.

After I observed several repetitions of this twitching, I requested her to try and move her feet. This, to the astonishment and gratification of all present, she in some measure accomplished, and was able to raise the toes of the left foot from the floor, and to put them down at pleasure; but the right foot and left heel were still immovable. She was then directed to move the toes of the left foot occasionally, or as often as convenient to herself, and after she had been in this semi-state a few hours she was requested to awake, with the resolution on her part that she would be able to use it when awake. She awoke, *and was able to move it as well as when asleep.*

One week after, I visited her again, and found her precisely as I had left her. *She was still able to raise the toes of the left foot, but nothing more.* She entered the state for the second time, and was directed to move the feet as before. This time she was able to move the toes of the right foot also, in the same manner as the left, but the heels of both were still beyond her control.

After being in the state several hours she was directed to awake as before, and when awake *was able to move the toes* of both feet. On the third sitting, she raised the left heel; and on the fourth, she gained control of the right. The fifth, she raised the left foot from the floor, and on the sixth, the right. She went on in this manner from week to week, *improving at every sitting, but never gained or lost anything between any of her sittings, although she made many efforts to move those parts which had not yet been relieved.*

By the middle of April she could stand, or lean the weight of her body upon her legs, and with the assistance of her friends was enabled to step or rather drag her legs after her.

About this time she was taken to her own home, about two miles and a half from the city, and I now visited her about twice a week. She continued to improve at every sitting, and by the middle of June, could walk alone; *her eyesight and general health also becoming better as her limbs improved.*

My visits were now again only made once a week, and, by the middle of July, I considered her *entirely restored.*

She, however, entered the state several times after, at intervals of two, three, and four weeks, at the same time she was able to follow her usual occupations.

This lady is also *one* of the very best clairvoyants, and I am indebted to her for many of the most important and interesting experiments, clairvoyance, hearing, feeling, tasting, and smelling, etc., which are detailed in this work.

I would here remark, that this lady did not become clairvoyant until after the twentieth sitting, although she, each time, made many efforts to become so.

The cause of her not seeing sooner I ascribe to her extremely nervous condition, and her not being clairvoyant prevented her from entering the state as deeply as I desired that she should, consequently her recovery was much slower than it otherwise would have been, had she been able to concentrate her mind more firmly, and to resolve more decidedly, that her disease should be forgotten, or cease to be, when she awoke.

XVI.—CASE.

About the middle of October, 1846, I was called upon by Mrs. ——, to visit her daughter, Miss ——, about eighteen years of age.

Upon my first visit I received the following general history of her case:

Her mother stated that she had always been very delicate, even from her childhood; that she had never been free from headache within her recollection, and that, for the last four years, she had been afflicted with a severe *cough* and a *continual pain in her chest.* For the last two years she had occasionally spit blood, and that during the last spring and summer she had expectorated considerable quantities of pus. That she had been confined to bed since last spring, and had had convulsions daily since that time, sometimes as many as *thirty* in twenty-four hours.

Soon after I entered the room she was seized with convulsions, and during the paroxysm, which lasted about three minutes, the whole body was convulsed, the muscles of the chest more particularly so; and, although conscious of their presence, she could not articulate, and to all appearance seemed to be in *great agony.*

After the spasms had subsided she seemed to be much exhausted, but soon after recovered sufficiently to converse as before.

She ascribed the convulsions—and I think correctly—to the *excessive pain* which she suffered, at intervals, in her "sides and breast."

She had consulted me as a last resort, and desired to know whether Somnambulism would have any effect upon her disease.

I stated, candidly, that her case was one which I

was fearful would not allow us to expect much from any remedy, but that if she would make up her mind to enter the state, I was convinced that she could, at least, be relieved of her pains and convulsions.

After I had given her the necessary instructions she made the attempt, but the convulsions coming on soon after, prevented our proceeding for that time.

She made another attempt in the evening of the same day, with but little better success.

Upon the third, however, she entered the state perfectly, and after waking felt much relieved.

She entered the state every evening for a week, and at the end of that time, the convulsions, and the pains in her head and sides, etc., had nearly subsided.

During the week, she had but three slight convulsions, and these were caused by local circumstances of a distressing nature. The family, though very respectable and intelligent, were in the most abject pecuniary circumstances. The father having died in the spring, of pulmonary consumption, the family, consisting of six small children, were dependent upon the mother and this her eldest daughter for support; but as she and several of the children had been ill since her father's death, they were unable to earn even the necessaries of life.

During this week she had also spit blood on two occasions, and expectorated considerable pus.

At the end of the second week the convulsions and pains in her head and chest, etc., had entirely subsided, she was able to sit up for several hours during the day, and there now seemed to be only a general weakness, and the original disease of her lungs.

About this time, as I entered the room one evening, I found her sitting up, and observed that she was spitting or throwing up something unusual, and upon inquiry, was informed that about half an hour after her meals, she has always thrown up what she has eaten, and had done so for the last twelve years.

This was the first intimation I had of this fact, and after she had been in the somnambulic state for an hour, just before waking I directed her mind to this circumstance, and requested her to resolve that this affection should also be corrected, and that her digestion should be improved. She made the resolution as I requested, and I had the satisfaction upon my next visit, of ascertaining that she had retained all the meals she had taken since my last visit.

After three or four other sittings, I found her appetite and digestion good, and her strength, spirits, and general appearance much improved.

About this time, in writing to a medical friend, after giving him a history of her case in detail, I concluded my letter by remarking: "What lying in the insensible condition of Artificial Somnambulism eight or ten hours a day may do for her lungs, I am anxious to know. The only question with me is, Will the perfect absence of irritation in the lungs for the above or a longer period, change the character of her disease, and induce the ulcerated surface to heal?"

Still further to show my views of her case at that time, I extract the following from the same letter:

"Heretofore, everything else has failed to give her relief, and as she enters the somnambulic state very deeply, and has been improving under instructions while in it, I have resolved to give it a fair trial, although my medical knowledge predicts unfavorably, and scarcely leaves room for any hope of permanent benefit. Time will show the result."

From the middle of November until the first of December she entered the state as usual every other evening, but was each time requested to awake, *with the exception of her body, particularly her chest.*

This she effected very readily, and soon learned to keep the chest and lungs in the insensible condition during the time between her sittings.

Under this treatment her cough and the expectoration of pus rapidly subsided.

From the first of December until the first of January, 1847, she entered the state but once a week, and is now—at the above date—so far as external appearances and her own feelings can warrant, entirely restored.

Within the last month she has been to church repeatedly, has visited her friends, and followed her usual occupations.

I lost sight of her from that time, and, upon inquiry, have since been informed that two years after she was restored by Somnambulism, she contracted a severe cold, which being neglected, in consequence of religious prejudices excited against Somnambulism by officious bigots, brought on the old condition of her lungs, and that she died of that disease three years afterwards.

The responsibility will of course rest where it belongs.

Ever since the result of the first trial with Somnambulism upon cases of pulmonary disease, I felt exceedingly anxious to continue the practice in others, but have as yet not been able to get cases sufficiently marked, or patients who are willing to make the trial—the facts in the case not being generally known—and will therefore have to await circumstances.

The foregoing cases, given in detail, have been selected from a number of others, who have been restored to health by the proper direction of the mind while in this state, within the last twenty-five years; and I will here again remark that the *mere entering this state will not relieve disease. It requires that the mind of the patient, while in this condition, should be directed to the disease, and a desire or a resolution formed on their part that it shall be otherwise when they awake. It is no matter whether this resolution be taken or be made independent of the instructor or not, the effect will be the same; but it is the duty of every person into whose care they entrust themselves, to see that it is properly done before they awake, or no beneficial effects will follow.*

Great care should also be taken that they do not imagine, resolve, think, fear, or believe that they will feel ill or badly when they awake, or this will certainly be the case. I have seen these effects upon many occasions, even in healthy persons who feared or conceited that they would be so. This should always be guarded against, and the current of their thoughts diverted from imaginary evils to pleasant scenes, or things that yield them joy instead of grief.

CHAPTER XXVI.

SURGICAL OPERATIONS.

I DEEM it nnnecessary to give in detail an account of the various operations which I have performed upon persons while in a state of Artificial Somnambulism, and will briefly remark that I have performed many which, though not the most important, are generally considered the most painful, without the patients feeling or even knowing that an operation had been performed upon them at all.

Among the number were several for cataract, the removal of tonsils, the tedious and difficult operation of removing a tumor from the inside of the mouth, and of the same from other parts of the body.

The removal by dissection of several great toe-nails, which had grown into the flesh on either side, and had been the cause of much suffering for years. The removal by an operation of a portion of glass which had penetrated deeply into the fleshy part of the hand; also in another case of a needle in the same situation, together with the extraction of innumerable teeth from various persons, who are all willing at any time to testify to the truth of what I have stated, and to acknowledge their preference for this method of getting rid of unruly members, or of passing through operations, etc., which are not to be avoided.

I shall conclude my remarks upon this interesting subject by stating that, in operations upon subjects while in this condition, *it is not only beneficial, because the patient is not subjected to the pain usually experienced while under severe operations, but because the system under such circumstances receives no shock, the effects of which every surgeon is fully aware, is more to be dreaded in their recovery than anything else.*

It is, therefore, self-evident, that when a patient has passed through an operation without pain or a shock to his nervous system, that his recovery must be more *sudden, pleasant, and certain*, than when he has not only suffered the pain and the shock, but must still necessarily feel the consequent irritation, etc., resulting on all such occasions.

CHAPTER XXVII.

OBSTETRICAL CASES.

IN the "*Boston Medical and Surgical Journal*" for October, 1846, vol. xxxv., No. 10, I published the case of a lady, whom I delivered of a full-grown healthy child, while she was in a state of Artificial Somnambulism, without feeling pain or interfering with the natural contractions of the uterus.

To this work I now refer all those who feel interested in the science or the benefits which it is destined to confer.

Two years prior to the publication of the case above referred to, I delivered a lady while in the same condition. The facts in her case are as follows:

In January, 1844, I, for the first time, delivered a lady of a healthy child, while she was in a state of Artificial Somnambulism, *without any pain to the mother*, except, when, for the sake of experiment, she was requested *to feel one pain and not another*.

I had, for a long time, been anxious to test the truth of my inferences, which were founded upon the facts laid down in this work, when speaking of the sense of feeling, where the powers of the subject to feel or not in any part as they please, is proved beyond dispute.

In the cases above alluded to the same is the case, and females have the same power over uterine pains, and, although they may feel a pain when they please they can also *not feel it when they are so disposed without interfering in the smallest degree with the natural contractions of the uterus, in expelling its contents.*

The fact, therefore, that woman can pass through the various stages of labor without feeling pain, has now been practically demonstrated.

This desideratum, which I had long believed possible, I have, with but little difficulty, accomplished in many other cases, and the time is not far distant when prejudice, ignorance, and bigotry will be set aside, and the benefits which an improved science has brought to our doors, will be hailed with delight by a free and enlightened people, while all the ills that flesh has been heir to will live but in the memory of the past.

This idea may seem premature at this time, but if mankind had witnessed the perfect freedom from pain, *in these cases*, during the severest contractions of the uterus, while in the act of expelling the fœtus from its cavity, they, too, would say:—" Old things have passed away, and a new era is at hand."

CONCLUSION.

Those who have read the preceding pages of this work carefully and without prejudice, I think must have arrived at conclusions similar to those which I have, from the deepest convictions of their truth, thought it my duty to advocate.

When this subject first came under my notice I found it literally to be a mass of contradictions and incongruities, over which ignorance, prejudice, and superstition had thrown a veil so dark and impenetrable that it was almost impossible to tell whether there was any truth commingled with the falsehood. And it is painful still to see, in our larger cities, the impositions of paralyzing limbs, and the seeing, hearing and tasting *falsely*, etc., permitted by the public, and practised by those who ought to know better, professing, as they do, to be the leaders of the science.

Those who have repeated the experiments which I have detailed in this work, will have seen that the *so-called operator* has no power to produce such effects; that the subject is independent of him in every sense of the word, and can see, move, hear, taste, feel, and smell, etc., independent of him or any other person, and that there is absolutely no more sympathy, in any sense of the word, between him and the subject, than there is between the subject and any person else.

That sympathy, of which so much has lately been said, only means fellow-feeling, and that no indescribable influence can be attached to it any more than there can be to the Neurara, or the animal magnetic fluid, neither of which have any existence in nature.

Nor is there any necessity for such fluids, or for such an influence, since every part of the system is absolutely and positively connected by nerves emanating from the brain, medulla oblongata and spinal marrow, and when we communicate with each other we use our sight, speech, hearing, feeling, taste, and smell, etc., all of which senses, while in a somnambulic state, can be translated to a distance—and this, too, whether we are conscious or not—and there used as well as if the persons or things were within our reach.

These fluids or influences, therefore, are only in the "mind's eye" of those who form their theories first, and afterwards endeavor to mould the facts to them.

I also deny that it is possible to put any person into this state or to take them out of it when contrary to their own will. Let the experiment be fairly tried, and it will be found impossible, because they do and always have done it themselves.

By a fair trial I mean, take a new subject that has never been trained to any system, and explain to him the true nature of the state and his powers therein, viz.: that he has a mind and will of his own, and, if he chooses, can use it in all conditions and under all circumstances, and it will be found impossible as long as he uses his will, for any one to throw him into any state, or into any condition to make him see, hear, do, taste, or smell, etc., anything contrary to his will, But if he be taught and *made to believe* that *white* is *black, black it will be to him in any condition; but because he believes so* does not prove that if the truth were taught him that he could not see its true color, or do as his reason dictates in spite of any one.

I have tried to effect the sleep hundreds of times, independent of the subject's knowledge, but could never succeed, unless the subject, or certain organs of his brain were in a clairvoyant condition; but I have frequently seen subjects fall into the sleep when *they thought* that I was willing for them to go into it, although I, at the time, had not the slightest idea of the kind.

If it were possible to *do this at all*, as has been said, it would, in my opinion, *be just as easy* to put any person into it, even a stranger at a distance, contrary to his will, and it ought to make no difference whether that stranger were in New Orleans or China.

The advocates of the sympathetic theory say that, " Where the *relation* has been sufficiently established between two persons a patient may be put to sleep as well a million of miles distant as one, provided he be in a suitable condition at the time, and have the necessary *apprehension* of the anticipated or *designed* result."

This is synonymous with, or, in other words, is as much as to say, if the subject be clairvoyant at the time and his attention be devoted to the so-called operator, that the result will follow. *This* is true. But my theory is, that the patient, even in this case, can throw himself into the state, not only independent " of the necessary *appreciation* of the anticipated design" of the operator, but in spite of him and contrary to his express will, either before his face or behind his back, etc., whenever he pleases. Let the experiment be properly made, and the truth of what I have stated will soon be rendered self-evident.

If there be an outside influence on an animal magnetic fluid in the case at all, it ought to exhibit its powers in all cases alike, as magnetism or any other imponderable fluid. Magnetism and electricity effects every person alike, and require conductors or wires when communication is to be established between persons at a distance.

I have, so far, failed to see the wires in the case of animal magnetism. Clairvoyance, and the powers of the senses, mind reading, etc., explain all the phenomena exhibited by persons while in this state, and I do not see why we should seek for a thing which, in reality, has no existence in nature, to explain phenomena which are already so perfectly accounted for by powers within ourselves.*

I think I have also sufficiently proved that the old method of exciting the organs, either by the application of the fingers of the operator, or of the subject's,

---

* The word clairvoyance does not express the idea I wish to convey when I speak of the powers of the senses, or of the mind, and, therefore, have used the word clearminded or clearmindedness, and wish it to be understood as meaning the power of the mind, or of all the senses and faculties. Clairvoyance is internal perception, or simply *seeing* without the aid of the eye.

is an incorrect way of obtaining information, and all that is absolutely necessary to produce the same effect is for subjects to throw their own minds upon those portions of the brain which may be designated to render them active.

In either case it is an act of their own will, and if they do not choose that an organ shall become active, their own fingers, or those of the operator, may be held there for an indefinite period, without producing any effect.

The only true method of ascertaining the location, etc., of the various organs is, in my opinion, to request the subjects to awaken—when the brain is in the somnambulic state—certain portions of the brain at a time; or when the brain is awake, to get them to put certain portions into this state, as I have fully described in this work.

When but a single portion of the brain has been awakened, and *kept perfectly awake*, it can be ascertained what powers of perception, etc., have been lost; and on the other hand, when but *a single organ is in this state*, it will be as easily ascertained what perceptions, etc., are the most or alone active.

That it is difficult, however, to get some subjects to do this properly, I have already acknowledged, but some do it very easily, and the results, I think, when proper care has been taken, will be *nearer the truth* than any other demonstrations which have yet been offered.

So far as my observations have gone, they have

been so, and although it requires a *great deal of care*, on the part of the subject, to keep a certain portion only either awake or asleep, and as it is the only way that will be likely to bring us nearer the truth, it should be steadily persevered in, until the accumulation of facts shall warrant our decisions.

With respect to the functions spoken of under the head of "The functions considered in the natural and somnambulic states," I have only to say that I believe that such kinds of action are possessed by each of the faculties, and that each faculty can act independent of the rest, producing ideas, the result of the action of the functions belonging to each; and that by their respective functions of association they are capable of acting together, and thus produce the various modifications of mind.

Much has been said in some works about subjects being agreeably or disagreeably effected by the touch of other persons, or other substances while in this state. They are, however, only likes or dislikes, either natural or from some causes, or indeed often only freaks which are exercised and changed by them at pleasure.

They may be insensible of pain when violence is applied to their own persons, yet at the same time, if their attention be directed to the "operator," or any other person, *whether they are suffering pain or not, they can imagine and feel pain*, as well in the one case as in the other; yet they are not *obliged* to feel it in either case.

This, as I have before stated, proves *that they can feel what they imagine as well as see what they imagine.*

If it were not so, they could not feel pain where none was felt by the "operator," or those who were engaged in the experiments. Their feeling pain in the above way cannot be ascribed to sympathy, unless we could suppose that he sympathized with himself, for no one else suffered.

So far as my experience goes, I have found that their powers of controlling the functions of the heart, kidneys, liver, etc., are very limited; but they can by an act of their own will, control *them just as well* without the application of the "operator's" fingers to certain parts of the head, or the supposed "sympathetic points," as with such an interference. They can "sing, dance, laugh," or in a moment render themselves rigid, insensible, or exquisitely sensitive to pain, without any operation upon any of the *newly invented* "cerebral functions," simply by an act of their own will.

This power, clearmindedness, and that of their ability to read the mind of another, has deceived many "operators," and caused them to *invent and locate organs* which really have no existence.

Their remembering what has transpired while in this state when they awake, depends upon their own will or determination to do so; and if they do not wish to remember, they can blot it from their memory and know nothing about it when awake.

Some subjects can, if they desire it, forget a name,

a person, or a place, etc., when they awake; they will have forgotten the one or the other as perfectly as if they had never known the name, been acquainted with the person, or had any knowledge of the place, etc., until they re-enter the state and resolve that it shall be otherwise; or, if they please, they can forget for a certain specified time, and at the end of that period will remember again.

This peculiar power, as I have before stated, I have taken advantage of for the purpose of curing diseases, etc., and recommend it as the best method which I have yet employed. I will here again remark, that the firmer the resolution on their part, the sooner the disease, the habit, or the affection, etc., will be overcome.

Some enter the state much deeper than others, and I have not yet seen two who do so alike; there can, therefore, be no description given, or rule laid down, which will apply to every case.

Some twitch and start very much while entering it, and others but very little; some feel cold and others warm; but, generally speaking, the hands and feet of those who enter it deeply become very cold, and remain so for a considerable time. In one or two cases I saw the breathing very much effected, becoming very quick and laborious; but I have frequently observed that most subjects are effected like, or similar to, the person or persons whom they first saw in the state, and am, therefore, of the opinion that all *unnatural appearances and affections*, etc., are the results, *not of the condition itself*, but of what they *saw in others*, and *believed to be* necessary or unavoidable.

Of all the phenomena exhibited by persons in this state, I consider their ability to read the mind of another one of the most extraordinary faculties which they possess. Their physical insensibility, their powers of seeing, hearing, tasting, smelling, and feeling things, etc., at a distance are extraordinary powers; but to read an idea, a thought, a mental image, a something that is not tangible, or which can neither be seen, heard, tasted, or felt in a natural state, I conceive is, if possible, still greater than all. How they do it they cannot satisfactorily explain, but that they *have done so* and *can do so again*, I do not now pretend to doubt.

I do not wish to be understood, however, that I believe that *all* subjects always do it correctly. This is, by no means, the case; but I have tested this power frequently in good subjects, and after calling their attention to the nature of the experiment I desired to make, without saying a word which might lead them to suppose where I had thrown my mind, or what I was thinking about, I have had them to tell me precisely what was passing in my mind at the time.

In experiments of this kind I have usually thrown my mind to a certain place, or upon a certain person or thing. Sometimes I thought of the name of a person or thing, or of places and scenes at a distance, but I could, as yet, never get them to repeat after me (verbatim) sentences of any length; but whether this was from inability, or a disinclination on their part, I am unable to say, and it is yet to be learned how far cultivation will improve this power.

The identity of this state with many others, some of which I have mentioned in this work, I think is self-evident, and although prejudices and a dislike to depart from old opinions will, for a while, retard the proper investigations by the masses, truth will, in the end, prevail.

It seems to be evident that subjects can have no positive foreknowledge of *what is to happen*, nor see anything that is past independent of the imagination, or by impressions received from other persons or bodies.

They may see things, at any distance which transpire, long before the fact can be communicated by persons, letters, or even by telegraph; but this is not strictly foreseeing, it is only seeing that which is passing at the time.

So, likewise, they may learn the past from another person who is familiar with that which may have taken place long before they were born; but they cannot see or know the past unless it can be obtained in a way yet unexplained, from surrounding or inanimate bodies, which, it would appear, retain a something that enables subjects even to distinguish articles, etc., belonging to strangers as well as others, although a number of the said articles may be commingled and shaken up in a hat. I have but to say, in regard to this matter, that all our surroundings seem to be affected by our presence in some photographic or mysterious way that scientific researches have not yet satisfactorily explained. In regard to the possibility of entering this state I have but to say, that the doing it depends upon the individual who desires to do so, as well as upon the instructions given, and it will be more difficult for those whose characters and prejudices are formed than for those who are free from such hinderances. The young and unsophisticated will, therefore, be more likely to succeed, and when the condition is perfectly understood by the masses, and properly taught by those who profess to do so, there can be no doubt that all will be able to enjoy its benefits. It does not require the gift of prophecy to foretell that the time is not far distant when Artificial Somnambulism will be taught in all the schools, lyceums, seminaries, and institutions of learning. Then will this knowledge be truly appreciated, and the benefits to be derived from it realized by all who wish to escape the ills that ignorance is heir to. But to conclude:

The consequent good resulting from a proper management of persons, while in this state, has been fully treated; but its benefits to mankind does not stop here. It goes hand in hand with Phrenology and many of the other sciences, and is the only key to the true philosophy of mind.

That what I have said of it is true, I have not

the slightest shadow of a doubt. I am fully persuaded that it will stand the utmost scrutiny, and like virgin gold, the oftener it is smelted the purer it will come forth from the furnace of its examination.

It is now but barely sifted from the dross, and in the mantle of unblushing truth presented to the world. Let it but have that justice which is due, and time will show whether it shall be denied the title of a Science.

THE END.

www.ingramcontent.com/pod-product-compliance
Lightning Source LLC
Chambersburg PA
CBHW032054230426
43672CB00009B/1594